Telecommunications Liberalization on Two Sides of the Atlantic

Telecommunications Liberalization on Two Sides of the Atlantic

Martin Cave
Robert W. Crandall
editors

AEI-Brookings Joint Center for Regulatory Studies

WASHINGTON, D.C.

2001

Telecommunications Liberalization on Two Sides of the Atlantic may be ordered from:

Brookings Institution Press
1775 Massachusetts Avenue, N.W.
Washington, D.C. 20036
Tel.: 1-800-275-1447 or 1-202-797-6258
Fax: 202-797-6004
www.brookings.edu

Library of Congress Cataloging-in-Publication data

Telecommunications liberalization on two sides of the Atlantic / Martin Cave and Robert W. Crandall, editors.
    p. cm.
Includes bibliographical references and index.
  ISBN 0-8157-0231-0 (pbk. :   paper)
  1. Telecommunication—United States—Case studies. 2. Telecommunication—Canada—Case studies. 3. Telecommunication—Europe—Case studies. 4. Competition, International. I. Cave, Martin. II. Crandall, Robert W.
  HE7775 .O67 2001
  384'.041—dc21

9 8 7 6 5 4 3 2 1

Typeset in Berkeley

Composition by Cynthia Stock
Silver Spring, Maryland

Contents

Foreword

The United States, Canada, and the European Union have embarked on ambitious programs of liberalization of their telecommunications sectors. The United States led this effort by opening its long-distance market to competition in the 1970s, but it was the enactment of the Telecommunications Act of 1996 that forced open the market for local and intrastate telecommunications. Canada was somewhat slower, liberalizing its long-distance market in 1992 and opening local telecommunications to competition in 1997. Finally, the European Union ordered all member states to begin opening markets to competition on January 1, 1998.

It is perhaps a little early for definitive judgments about the success of these efforts, but the authors in this volume offer some tentative conclusions about the directions of reform on both sides of the Atlantic. Robert W. Crandall and Thomas W. Hazlett are mildly sanguine about the results in the United States, although they prefer the less regulatory approach adopted in Canada. Martin Cave and Luigi Prosperetti are less optimistic about the European Union. Chapters 2 and 3 critique individual country policies and suggest reforms that would speed the path to competition.

This volume is one in a series commissioned by the AEI-Brookings Joint Center for Regulatory Studies to contribute to the continuing debate over regulatory reform. The series addresses several fundamental issues in regulation, including the design of effective reforms, the impact of proposed reforms on the public, and the political and institutional forces that affect reform.

The debates over regulatory policy have often been highly partisan and ill-informed. We hope that this series will help illuminate many of the complex issues involved in designing and implementing regulation and regulatory reforms at all levels of government.

ROBERT W. HAHN
ROBERT E. LITAN
AEI-Brookings Joint Center for Regulatory Studies

1

Telecommunications Policy in North America and Europe

Martin Cave
Robert W. Crandall

The 1990s witnessed a major revolution in telecommunications policy in North America and Europe. Although telecommunications liberalization had begun in the United States in the 1970s and in the United Kingdom in the mid-1980s, there was no consensus on the need to substitute competition for private or public monopoly on either side of the Atlantic until recently. By the early 1990s, however, the electronics revolution had swept the world, and most countries began to realize that they could not compete in many markets without a vibrant, competitive telecommunications sector. As a result, the European Union, Canada, and the United States launched major new liberalization policies that are aimed at opening all telecommunications markets to competition. The United States had a clear first-mover advantage in some markets, but the EU and Canada are striving mightily to catch up. In this volume, we present two views of the progress toward competition—one for North America and one for Europe.

1

Market Structure before 1990

Most of the world's telecommunications systems were govern-ment owned until recently. As a result, most countries did not have a need for regulatory authorities to check the monopoly power of private telephone companies. The major exceptions were found in the United States and Canada.

North America. Neither Canada nor the United States had large government-owned telephone monopolies before the recent lib-eralization trend. The American Telephone and Telegraph Com-pany dominated the U.S. telecommunications sector until 1984, when it was broken up into a long-distance and manufacturing company, AT&T, and seven regional Bell operating companies that offered local service, limited-range long-distance service, and directory services. Long-distance service had been opened to competition in the 1970s, and this competition became much more intense after the 1984 divestiture. Local services, however, were generally offered by a single franchised monopolist in each area. These local monopolists included the operating compa-nies divested from AT&T in 1984, GTE, and several other inde-pendent companies and many small rural cooperatives.

In Canada, all services other than private line services were provided by vertically integrated monopolists until the early 1990s. Bell Canada had a virtual monopoly in Ontario and Quebec. Most of the other provinces had one large telephone company that offered local and long-distance services throughout most of the province. In a few provinces, such as Saskatchewan, the pro-vincial government owned the telephone company, but most access lines were controlled by regulated private companies.

In Canada and the United States, regulation of telecommuni-cations was in the hands of federal and state (provincial) regula-tors. In the United States, all interstate services were regulated by the Federal Communications Commission, but intrastate (in-cluding local) services were regulated by state commissions. In Canada, however, the national regulator—the Canadian Radio-Television and Telecommunications Commission—did not have the authority to regulate interprovincial services offered by government-owned telephone companies until 1993.

In Canada and the United States, telephone rates were enormously distorted by regulators for political ("universal service") reasons. Both countries required that local service be offered at very low flat monthly rates to residential subscribers, particularly in rural areas. Long-distance rates were kept far above cost, especially in Canada where long-distance competition did not exist before 1990.

Local business rates were also allowed to remain far above cost to cross-subsidize residential service, and neither country allowed competition in local markets even though both had highly developed, almost ubiquitous cable television service that could potentially be adapted to the delivery of local telephone service.

Europe. For the most part, European telecommunications companies were government-owned monopolies as late as 1990. The United Kingdom had privatized its national carrier in 1984 and allowed limited entry in 1985. A few countries followed suit, but the other EU countries had not begun to privatize their national monopolies in 1990, much less to admit competition.

In 1990 virtually every European country had greatly distorted telephone tariffs with low line-rental rates and high local and long-distance calling rates. Unlike the United States and Canada, European telephone companies exacted substantial charges for local calls that varied by time of day. These charges became a major obstacle for the development of the Internet later in the decade.

Regulatory Changes in the 1990s

Telecommunications policy underwent a revolution in the 1990s on both sides of the Atlantic. Major new legislation or regulatory edicts were passed in the United States, Canada, and the European Union.

North America. In the United States, the effects of the 1984 AT&T divestiture began to create pressure for a major change in policy. The regional Bell operating companies wanted to enter long-distance markets, and the long-distance companies were eager to enter local markets. The resulting 1996 Telecommunications Act provided both opportunities. All telecommunications markets were opened to competition, and incumbent local companies were required to interconnect with entrants by offering to lease portions or unbundled "elements" of their networks to these entrants.

In Canada, the 1993 Communications Act established that the Canadian Radio-Television and Telecommunications Commission had regulatory authority over all non-provincially-owned telephone companies. In 1992 the CRTC first admitted entry into switched long-distance services. Then in 1995, the CRTC extended its liberalization policy to local service, requiring incumbents to interconnect with local competitors but requiring much less network unbundling than did the United States.

Europe. Throughout the 1990s, the European Union began to build up a patchwork of regulatory principles to be implemented by national regulatory authorities for a liberalized telecommunications market. More countries followed in the footsteps of the early liberalizers, and the member states finally agreed to open their markets to infrastructure and service competition in 1998.[1] Significant privatization took place, but even by the end of the decade, the government continued to own shares in the major operators, such as France Télécom and Deutsche Telekom. Incumbents were required to negotiate interconnection agreements with entrants, but the commission's guidelines did not initially require unbundling of network elements. Universal service requirements were to be made explicit, and a regime for sharing universal service obligation (USO) costs was made available. However, enforcement of the legislation was patchy, and competition was slow to develop in some countries.

Changes since Liberalization

Given the enormous distortions in most countries' telephone rate structures before liberalization, opening telecommunications markets to competition might be expected to wreak havoc with existing tariff structures. In fact, the change in local telecom rates has been much more gradual because entry has been so slow to occur. Nevertheless, long-distance competition has developed on both sides of the Atlantic, creating major changes in customer rates.

North America. Long-distance competition began in the 1970s in the United States and in 1992 in Canada. In both countries, long-distance rates have fallen substantially, in part because regulators have slowly reduced carrier access charges. Today, Canada appears to have slightly lower rates than the United States.

Local competition is much more recent in both countries. It has taken much of the past five years to implement the provisions of the 1996 U.S. Telecommunications Act that guide local competition; hence, one cannot draw definitive conclusions about the effectiveness of the act's approach to providing incentives for entrants to use incumbents' facilities. However, Robert W. Crandall and Thomas W. Hazlett show that facilities-based competition is beginning to develop as the result of the new competitive local-exchange carriers' substantial investment in new facilities. The competitors have wrested about 7 percent of access lines from the incumbents and about 9 percent of revenues. Canada's local-competition policy was not enunciated until 1997, and local competition has been much slower to develop there because the cable television companies have been slow to move into telephony.

Europe. As of late 1999, interconnection agreements between incumbent telephone companies and entrants had been negotiated in all EU countries. The incumbents' share of the long-distance market declined substantially in Sweden, Germany,

Belgium, and the United Kingdom. Martin Cave and Luigi Prosperetti show that interconnection rates vary substantially across the EU. Although many rates are a multiple of the commission's suggested "best practices" rates, several countries exhibit extremely low rates.

Competition had begun before 1998 in five countries, but only the United Kingdom had new entrants with more than a combined 10 percent of access lines. As of 1999, only a few EU member states had adopted regimes for unbundling the local loop, but it was required everywhere by the end of 2000. Considerable controversy continues over the terms of interconnection among fixed wire operators and between fixed wire and mobile operators. Despite this turbulence, mobile penetration has grown rapidly and is higher than in the United States.

Cave and Prosperetti find that several national and pan-European companies are now entering telecommunications markets in Europe, but they argue that it is too early to characterize competition as effective in *most* markets. They are concerned that high interconnection and leased-line rates throughout Europe are restricting competition significantly.

New Technologies and the Impetus for Further Change

Much of the policy that drove liberalization on both sides of the Atlantic was based on a backward-looking assessment of the need for competition in telephony. Entry was to be promoted in basic voice-data services through the interconnection of traditional circuit-switched networks. Incumbents had to be reined in because they had a first-mover advantage that derived from their ubiquitous networks and trade names. Much has changed since the mid-1990s when these views were guiding the reform process.

The Internet is now driving all developments in telecommunications. It now seems likely that all or most consumer telecommunications will shift to the Internet. If this move occurs, the current regulatory regimes on both sides of the Atlantic will face severe challenges because all of them require the incumbent carriers to recover a substantial share of their costs from charges that are assessed on the basis of minutes of use. These tariff regimes, designed to cross-subsidize local connections in the name of universal service, cannot survive an Internet-dominated environment, especially in Europe, where large local charges have suppressed the use of the Internet already.

Europe is probably ahead of the United States and Canada in developing wireless as an alternative to wireline connections. The European calling-party-pays approach to mobile pricing and high local charges have led to greater diffusion of cell phones and, therefore, a greater potential for wireless-wireline substitution. As Cave and Prosperetti point out, European countries have auctioned UMTS (3G) spectrum far in advance of North America. Whether the high prices paid for this spectrum in the competitive UK and German auctions are an advanced indication of the potential for broadband wireless services remains to be seen.

Conclusions

Neither Europe nor North America can be said to have developed the ideal regime for developing competition in telecommunications markets, but valiant efforts are being made on both sides of the Atlantic. The United States and Canada have advanced competition in long-distance services rather successfully. The EU contains some countries in the early stages of liberalization and privatization and others that are much more advanced. The substitution of competition policy for regulatory policy would undoubtedly move the process of liberalization more rapidly on both continents, and the EU is making tentative steps in that direction.

2

Telecommunications Policy Reform in the United States and Canada

Robert W. Crandall
Thomas W. Hazlett

The regulation of telecommunications in the United States and Canada has undergone formidable reform in recent years. Most attention is understandably focused on the United States' Telecommunications Act of 1996 because it marked a fundamental departure from decades of established regulatory policy. The Telecommunications Act was an official declaration by the Congress of the United States that the basic assumptions of the 1934 Communications Act were defunct. Monopoly market structures are no longer presumptively efficient and best accommodated through common carrier rules and rate regulation. Instead, competition is now to be phased in because it is presumed to be the better market alternative for customers and for providing incentives for productivity growth. Barriers to entry in local and long-distance telephony, as well as cable television delivery service, are to be eliminated. Moreover, transitional mechanisms are to be used by regulators to squeeze competitors quickly into the marketplace.

By contrast, Canada has introduced competition in its telecommunications sector through administrative decisions of its regulatory authority, the Canadian Radio-Television and Telecommunications Commission (CRTC) under a new Telecommunications Act that allowed such liberalization but did not require it. The CRTC opened Canadian long-distance markets to competition in 1992, more than two decades after interstate competition began in the United States. Five years later, the CRTC also opened local markets to competition, but—as we shall show—in a less disruptive and contentious environment than that produced in the United States by the 1996 Telecommunications Act.

In this chapter we focus most heavily on the impacts of the new Telecommunications Act in the United States, comparing U.S. results with those from Canada whenever possible. The conventional wisdom, expressed regularly in a spate of "anniversary" articles appearing about February 8 each year,[1] is that telephone competition in the United States has lagged, while a string of megamergers has combined the largest telecom providers into market-dominating behemoths. Rates are alleged to have risen, not declined, as advertised by policymakers. Leaders of both political parties, in defense, continue to tout the bipartisan legislation, suggesting that more time is needed to observe the benefits yet to come. Curiously, rising cable TV rates and declining cellular telephone rates are often cited as outcomes of the act, even when the 1996 law had little or no impact on the policies that govern these services.

Sorting out the effects of legislation—pulling out the actual effects associated with one law when legal and economic changes are buffeting the sector—is yeoman's work. In this chapter we attempt a modest first step by examining broad trends within the U.S. telecommunications sector to discern if they are consistent with the announced goals of the legislation. No doubt more intensive and subtle study lies before us in the many years of telecommunications policy debate ahead.

Background

The regulation of telecommunications in the United States has always involved a complex struggle among the states, the Federal

Communications Commission (FCC), and the courts. Liberal-ization of long-distance services began when the FCC inadvert-ently allowed MCI to begin offering switched services in the mid-1970s. Failing to construct a serious argument why such entry should not be allowed, the FCC was essentially defense-less in the federal court system when MCI began offering switched interstate (long-distance) services without an FCC license.[2] The FCC scrambled to devise an access-charge policy, eventually set-tling on a system that provided entrants with subsidized access rates until an equal-access regime was implemented by local-exchange carriers as the result of the 1984 AT&T divestiture. Eager to protect competitors thereafter, the FCC developed a "dominant-carrier" regulatory regime through which it restrained AT&T's competitive impulses through 1995.

Until 1996, virtually all liberalization in the United States was undertaken by the FCC or (to a much more limited extent) by state regulatory commissions without federal legislation. Al-though a variety of court cases had shaped U.S. telecommunica-tions regulation over the previous sixty years, the 1934 Communications Act's basic framework for the regulation of wireline telecommunications services remained largely un-changed. States had the authority to regulate intrastate wireline services and, therefore, to block entry into the delivery of these services. Most accepted the invitation eagerly until 1996.[3] For example, no state moved before 1996 to require equal carrier access for intrastate long-distance calls. Only six states allowed even a modicum of local-service competition for dispersed small business and residential customers, although most had allowed competitive urban fiber-optics rings to be built in large cities by competitive access providers (CAPs). This equilibrium might have remained undisturbed, but for the suffocating effects of the AT&T divestiture on the regional Bell operating companies (RBOCs).

In 1984, seven RBOCs were established to take over the lo-cal-exchange operations of AT&T. These new companies were expressly prohibited from manufacturing equipment, offering "information services," or offering long-distance services outside their local access and transport areas (LATAs).[4] These provisions of the AT&T divestiture decree prevented the divested Bell companies from participating in the national long-distance market and greatly frustrated their attempts to develop new services and technologies.[5] After failing to obtain relief from the court enforcing the AT&T decree, the RBOCs turned to a legislative solution. In 1996, they finally obtained such legislation, but the price was high: a new asymmetric regulatory regime and the liberalization of entry into local and intrastate markets.

The U.S. Telecommunications Act of 1996

At least twenty years in the making, the Telecommunications Act of 1996[6] is the most comprehensive piece of U.S. legislation to be enacted in this sector since the 1934 Communications Act. Its stated purpose is "to provide for a pro-competitive, de-regulatory national policy framework designed to accelerate rapidly private sector deployment of advanced telecommunications and information technologies and services to all Americans by opening all telecommunications markets to competition."[7] One hundred and twenty-eight pages later (Adobe Acrobat version), the rules are laid out defining how this transition is to proceed.

Despite its "de-regulatory" purpose, the act mandates an extraordinary number of regulatory proceedings to be conducted by the FCC. Pursuant to the act, the commission was to conduct more than eighty separate rulemakings or investigations. By October 1997, the FCC had listed some 184 Reports, Orders, Public Notices, meetings, or hearings associated with FCC responsibilities under the act.[8]

While the act addresses a sweeping range of activities, the big-ticket economic items involve the deregulation of three markets:

—Long-distance telephone service;
—Local-exchange telephone service; and
—Local cable television service.

In each of these areas, the act is designed to increase competition. In long-distance (IXC) service, the basic policy reform is to permit the RBOCs to enter subject to certain conditions. The most important are the obligations to obtain state certification from public utility commission (PUC) regulators that the company has opened its market to local competition; FCC and Department of Justice certification that a fourteen-point checklist of requirements is satisfied, guaranteeing that local markets are open to competitors; and FCC certification that RBOC entry into long distance would be in the public interest.

The act places these requirements on the Bell companies to accelerate local-market competition; the RBOCs are not allowed to enter lucrative long-distance markets until they facilitate entry into their own local markets. But two additional important reforms are applied specifically to local markets. First, the act requires that "no State or local statute or regulation, or other State or local legal requirement, may prohibit or have the effect of prohibiting the ability of any entity to provide any interstate or intrastate telecommunications service."[9] This eliminated the monopoly franchises for local telephone service issued by most U.S. states. Second, the act mandates that the incumbent carriers provide "unbundled access" to their networks for any entrant that wishes to use parts of their networks. The incumbent carriers have "the duty to provide, to any requesting telecommunications carrier for the provision of telecommunications service, nondiscriminatory access to network elements on an unbundled basis at any technically feasible point on rates, terms, and conditions that are just, reasonable, and nondiscriminatory." Besides the unbundling obligation, the act requires incumbents to allow entrants to resell their retail services by allowing the entrants to buy the entire package of customer services at a wholesale discount.[10]

Cable television markets are subjected to a phase-out of rate regulation and the abolition of restrictions on telephone company provision of video services. Developments in these markets will not be covered in this chapter.[11] Nor will the myriad other issues touched on in the act, including the V-chip mandate (program ratings for sex and violence on television), relaxation of limits on radio and TV station ownership, pre-emption of any auction for digital television licenses, the Communications Decency Act (restricting online content deemed harmful to children), utility pole attachment rules, and reform of universal service subsidies.[12] Our chapter focuses on how the act's major economic initiatives in local and long-distance telephony have succeeded in delivering benefits to consumers. We compare the progress in the United States with that achieved in Canada under its more recent liberalization policies. Finally, we include a section detailing some basic public choice issues regarding the U.S. Telecommunications Act.

Long-Distance Competition in the United States and Canada

It is commonly assumed that competition in the U.S. long-distance market could not develop until the courts dismembered AT&T, separating the local "bottleneck" facilities from the long-distance and manufacturing operations. The 1982 consent decree divesting AT&T was designed to remedy unlawful acts in the 1970s to restrain competitive entry into the long-distance and terminal-equipment markets. But vertical divestiture—though arguably a sufficient condition for establishing competition—was not a necessary condition. For this reason, the CRTC and the Canadian competition authorities have not attempted to force vertical divestiture on the Canadian incumbents—Bell Canada, Telus, BC Tel (now part of Telus), MT&T, Island Telephone, and NewTel. All continue to offer local and long-distance service.

In the United States the divested Bell operating companies had been barred from the long-distance market for twelve years before the passage of the 1996 Act. The new law allows them to enter this market on a state-by-state basis, but only after three regulatory authorities—the state regulatory commission, the Department of Justice, and the FCC—certify that the Bell company is in compliance with the act's interconnection requirements. In Canada there is no such quid pro quo.

The Canadian approach to facilitating entry relies on simple interconnection between networks rather than on interconnection for resellers or lessees of incumbent facilities. In the case of long-distance services, the CRTC learned from a critical U.S. error—the failure of regulators to mandate equal access to local switches.[13] The FCC had such an opportunity in 1969 (when MCI was first allowed to enter as a private-line carrier), in 1971 (when private line entry was allowed generally), and in 1977 (when the courts pried open all long-distance services), but declined to take it. Equal access for all long-distance carriers became a reality only when it was mandated by the 1982 decree that broke up AT&T and was subsequently extended to non-Bell local companies by the FCC. Compliance was generally not achieved until 1986–87, more than a decade after MCI began offering ordinary (switched) long-distance service. By contrast, the CRTC required that incumbent carriers provide equal access to all certified entrants in its 1992 order opening the long-distance market to competition. As we shall see, this requirement would quickly unleash long-distance competition.

Market Results. The United States began to admit competition into long-distance services more than twenty-five years ago. Canada began much more recently, waiting until 1992 to allow facilities-based competition. Nevertheless, Canada's long-distance market is now at least as competitive as that of the United States. The U.S. long-distance market has become much less concentrated since MCI ventured forth in the mid-1970s. Because U.S. local carriers were not required to offer equal access to long-distance carriers until AT&T was broken up by the courts, most analyses of U.S. long-distance competition begin with 1984.

Market Concentration. Long-distance carriers may be facilities based or resellers. Since 1984, the number of both types of U.S. carriers has expanded dramatically. Many of these carriers offer services over only a small region, but the number of national carriers has grown steadily. Between 1984 and 1995, AT&T was a regulated carrier whose pricing discretion was limited by a

Table 2-1. Market Shares of U.S. Long-Distance Carriers, 1984–99[a]

Percentage of total revenues

Year	AT&T	MCI	Sprint	WorldCom	Others
1984	90.1	4.5	2.7	. . .	2.6
1985	86.3	5.5	2.6	. . .	5.6
1986	81.9	7.6	4.3	. . .	6.3
1987	78.6	8.8	5.8	. . .	6.8
1988	74.6	10.3	7.2	. . .	8.0
1989	67.5	12.1	8.4	0.2	11.8
1990	65.0	14.2	9.7	0.3	10.8
1991	63.2	15.2	9.9	0.5	11.3
1992	60.8	16.7	9.7	1.4	11.5
1993	58.1	17.8	10.0	1.9	12.3
1994	55.2	17.4	10.1	3.3	14.0
1995	51.8	19.7	9.8	4.9	13.8
1996	47.9	20.0	9.7	5.5	17.0
1997	44.5	19.4	9.7	6.7	19.6
1998	43.1	[Acquired by	10.5	25.6	20.8
1999	40.7	Worldcom]	9.8	23.7	25.7

Source: FCC (2001).

a. Excludes local-exchange carriers' long-distance revenues, but includes both intrastate and interstate revenues of long-distance carriers.

price-cap regime with rate floors and by a requirement to file tariffs. To some observers, this combination of price caps and tariff-filing requirements provided a convenient mechanism for tacit collusion among the three largest carriers, AT&T, MCI, and Sprint.[14] Shortly after the imposition of FCC price caps, MCI and Sprint's market share growth began to slow down (table 2-1). Between 1986 and 1991, MCI and Sprint's combined market share of national long-distance revenues grew from 11.9 percent to 25.1 percent. Since 1991, it has grown by only another 8 percentage points to 33.5 percent, even with MCI's merger with WorldCom. A Herfindahl-Hirschman Index (HHI) based on total long-distance carriers' revenues[15] continues to decline, but it

Table 2-2. Canadian Long-Distance Market Shares, 1995–99

Percentage of minutes

Company	1995	1996	1997	1998	1999 (Q2)
Former Stentor (incumbent) companies	78	71	66	64	65
AT&T Canada	8	11	12	10	. . .
Sprint Canada	8	11	14	12	. . .
Others	6	6	7	14	. . .
Total nonincumbents	22	28	33	36	35

Sources: 1995: Stentor Hearings evidence in Canadian Radio-Television and Tele-communications Commission Forbearance Proceeding;1996–98: Call-Net interrogatory responses in CRTC 99-5. Yankee Group, Canada, August 3, 1999.

remains at more than 2500, far above the threshold for Department of Justice action on horizontal mergers.[16]

Although Canada's liberalization occurred nearly two decades after MCI began offering switched long-distance service in the United States and eight years after the AT&T divestiture, long-distance competition in Canada is well advanced. Spared from the contentious court debates that clouded the U.S. environment and proceeding much more deliberately in implementing equal access, the Canadians have avoided much of the transition required in the United States to move from monopoly to a more competitive market. Indeed, because Canada did not pursue vertical divestiture, the incumbent local companies are aggressive competitors with a shadow price of access that is equal to marginal cost.

Within six years of Canada's long-distance decision, the incumbent companies had lost about 35 percent of their market shares (table 2-2). In the United States, AT&T's market share fell from 84 percent of interstate minutes in the third quarter of 1984 to 65 percent in 1989, five years after divestiture and about fourteen years after MCI began to offer switched long-distance service. These results suggest that an equal-access regime without divestiture can work well to ensure entry into long-distance services.

Prices. Because long-distance rivalry among the largest carriers has taken the form of intense marketing of a bewildering array of discount programs, it is difficult to measure the degree to which long-distance rates have declined with regulated access charges and other costs. William E. Taylor and Lester D. Taylor contend that for much of the decade after divestiture, AT&T's rates fell by less than its access costs.[17] Robert Hall has argued that their analysis is misleading because it fails to take discount plans into account.[18] Paul W. MacAvoy contends that even when one allows for discounts, the price-cost margin in long-distance services has risen since 1990. Moreover, MacAvoy shows that a substantial number of telephone subscribers do not avail themselves of discount pricing plans.[19]

Ultimately, any judgment about the degree of competition is based on the proximity of rates to incremental cost. Figure 2-1 shows the trend in average long-distance prices in the United States and Canada since 1992, the year in which the CRTC opened the Canadian market to competition. Access charges are now similar in Canada and the United States, yet by 1998 Canadian long-distance rates had fallen below those in the United States. Equal access and the ability of the incumbent local carriers to compete aggressively appear to be sufficient to generate results that now surpass those in the United States more than twenty years after MCI began offering switched long-distance service. This result strongly suggests that it was not vertical divestiture, but equal access, that created the environment for long-distance competition.

The rates shown in figure 2-1 are averages for all long-distance customers, business and residential. Average residential rates are much higher for U.S. consumers. In 1996–97 interstate rates averaged about 17.5 cents per minute; in 1998 they were 15.3 cents per minute.[20] The recent introduction of new 5 to 7 cents-per-minute plans reduced the average consumer charge to 14 cents per minute in 1999—fifteen years after divestiture. In Canada, just seven years after the introduction of switched long-distance competition, carriers offer residences off-peak rates of as little as 1.6 cents (U.S.$) per minute.[21]

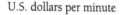

Figure 2-1. Average Domestic Long-Distance Rates in Canada and the United States

U.S. dollars per minute

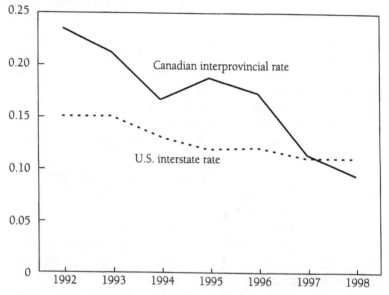

Sources: FCC (2000, table 9); for Canada, Midland Walwyn; Canadian Radio-Television and Telecommunications Commission, Docket 99-S, 1999.

Through 1998, long-distance rates in both countries were substantially in excess of long-run incremental cost. The long-run incremental cost of long-distance service is estimated to be no more than 2 cents exclusive of marketing costs and administrative overhead.[22] Wholesale rates in Canada are now about 2.5 cents (U.S.$) per minute, and they have been as low as 1.5 to 2 cents per minute in the United States. Even with access charges at 5 cents per minute, rates should decline to substantially less than 10 cents per minute in a competitive market for customers with heavy usage.[23] Indeed, business rates in both countries in 1998 averaged about 8 cents (U.S.$) per minute. Given current

U.S. carrier access charges of 2 cents per conversation minute, the long-run incremental cost of residential service, exclusive of marketing and other administrative costs, is already less than 5 cents per minute, equal to the lowest rate now available.

When the Bell companies are finally admitted into the long-distance market in all states, they are likely to combine in-region facilities with resale to compete in the inter-LATA market.[24] They should be able to lure a substantial share of their own in-region local subscribers to their long-distance service. This entry, in turn, could set off a much more vigorous general round of long-distance price reductions that could even reach residential subscribers who were still paying an average of 14 cents per minute in 1999.[25]

The losses to consumers from barring Bell entry into long distance have been substantial. Had long-distance rates fallen as rapidly in the United States as in Canada—a reasonable counterfactual given that vertically integrated Canadian incumbents continued to compete in long distance—rates might have been substantially lower throughout the 1990s. At 1998 volumes, the annual increase in consumer welfare for a 1.1 cent per minute reduction in interstate residential rates is $1.2 billion.[26] Were intrastate rates also reduced by a similar amount, the consumer welfare gains would be about $1.8 billion per year. Since business calling volumes are approximately twice the residential volumes, the consumer welfare gains from lower prices of telecommunications-intensive final goods and services would surely be a multiple of this $1.8 billion per year, depending on the derived-demand price elasticities. Had Bell-company entry into long-distance services been permitted in the 1990s, interstate long-distance rates would certainly have been several cents per minute lower, and consumers would have realized welfare gains that would be reckoned in the tens of billions of dollars. Blocking competition—ostensibly to promote competition in other markets—has incurred substantial social costs. These losses are reminiscent of earlier FCC decisions delaying entry into cellular and voice-mail services.[27]

Local-Exchange Competition in
the United States and Canada

It would be an understatement to suggest that there was skepticism about the feasibility of local competition during the debate leading up to the 1996 U.S. Telecommunications Act that "deregulated" the telephone business. The popular view of the local telephone exchange was one of "natural monopoly." The established local carriers—the RBOCs and GTE—were viewed as having enormous first-mover advantages that Sprint, MCI, AT&T, or the cable television companies cannot easily overcome. Regulators therefore attempted to jump-start local competition through a variety of policy instruments beyond (legally) open entry.

Before 1996 few states had competitive local carriers except for the fiber-ring competitive access providers (CAPs) in urban business centers. Dispersed small businesses and residential subscribers had no alternative to the traditional regulated local-exchange carriers (LECs) for wireline local service. Large businesses could utilize PBXs, purchase direct connections to long-distance carriers, or develop their own private networks and thereby avoid many of the above-cost rates that were forced on smaller businesses to cross-subsidize residential services. These choices were not available to smaller subscribers.

The 1996 Act requires state regulators to admit entrants into the provision of local and intrastate services. The first step in the process of gaining entry, however, is negotiation of interconnection agreements with incumbents. These interconnection agreements specify, among other provisions, the rates for unbundled elements, the rates for exchanging traffic, points of interconnection, and wholesale discounts. Under the 1996 Act, the incumbents must unbundle their networks into separate facilities or "elements" and lease them to entrants. This requirement has been interpreted by the FCC to require that virtually all incumbent facilities be unbundled except for the new equipment used to deliver broadband services.

Table 2-3. The Distribution of RBOC Unbundled Business Loop Rates (UNEs) and Wholesale Discounts for Business Service, Lower 48 States, 1998

Rate and discount	Number of states	Share of access lines (percent)
Unbundled business loop rate— most dense area (dollars per month)		
Less than 10	6	20.8
10–14.99	15	48.0
15–19.99	16	22.2
20–24.99	6	5.5
25 or more	4	3.6
Wholesale discount for business service (percent)		
Less than 15	9	6.2
15–17.49	11	0.4
17.50–19.99	14	32.0
20–22.49	12	22.8
22.5 or more	2	3.7

Source: Crandall and Hausman (2000), derived from industry sources.

The FCC has also ruled that the rates established in state arbitrations should be grounded on a forward-looking measure of long-run incremental cost, total element long-run incremental cost (TELRIC). Under this approach, the entrants may lease any facilities they choose for any period of time at TELRIC rates that assume the facilities are held for their full useful economic lives.[28] This approach, providing a "free option" to entrants, is still the subject of federal litigation.[29]

The rates that have resulted from these arbitrations/agreements are shown in table 2-3. Note the enormous variance in these rates despite the FCC's ruling that TELRIC would be the model by which the states would determine the cost of local access. The data shown in table 2-3 are for the densest areas in each state where states "de-average" these rates.[30] The forward-looking cost of serving these areas should be relatively similar across states, but the state-arbitrated rates widely vary. The politics of rate setting apparently prevent FCC methodology from consistent application.

Similarly, local and intrastate long-distance rates reflect the redistributive politics that drive state regulatory actions. Local rates for businesses are far above local residential rates, and state regulators have not allowed these rates to change very much since 1996 (table 2-4). Intrastate long-distance rates also vary enormously, with rural states having much higher rates in order to cover their proportionately larger deficits from setting rural residential rates far below cost.[31]

Canada's local competition policy does not require interconnection at all "technically-feasible" points, as required by the U.S. 1996 Act. Instead, it simply requires that each LEC must designate or establish a local point where traffic is exchanged. The existing wire centers are where interconnection takes place.

Only "essential" facilities, such as local loops in rural areas, must be unbundled and then only for a limited number of years. Resale is permitted, but incumbents do not have to offer resale discounts to entrants. Nor does Canada use the "carrot" of allowing the incumbent local companies into long distance as an inducement to facilitate this interconnection. The local companies are already in the long-distance business. In short, Canada's local competition policy is far less regulatory, relying more on facilities-based competition than on entrants' use of incumbent facilities.

Market Results for the United States. Despite the problems of regulation-intensive policy, entry into local telephony by new competitors is now occurring in the United States. Whether such entry is due to the elaborate TELRIC rules for unbundling or the simple prohibition of monopoly franchises by state authorities is the subject of intense debate. A principal criticism of the 1996 Act's requirements for comprehensive unbundling and the FCC's TELRIC pricing policy is that they combine to reduce the incentive for entrants to build their own facilities. In addition,

Table 2-4. Monthly Residential and Single-Line Business Rates in Selected Cities, October 15, 1994–98

Dollars per month

City	1994	1995	1996	1997	1998
			Residential rates		
Pine Bluff (Ark.)	22.22	22.06	22.14	22.22	22.22
San Diego (Calif.)	12.18	15.59	15.69	15.57	16.01
Atlanta (Ga.)	24.53	24.80	24.98	24.98	24.92
Chicago (Ill.)	18.20	17.31	17.63	17.18	17.18
Louisville (Ky.)	24.17	23.66	23.66	24.63	24.63
Baltimore (Md.)	24.98	24.98	24.98	24.98	24.67
Boston (Mass.)	23.07	23.07	23.07	23.07	23.07
Grand Rapids (Mich.)	17.53	18.06	17.95	18.01	18.25
Butte (Mont.)	18.22	18.22	18.22	19.26	19.69
Memphis (Tenn.)	20.25	20.25	20.33	20.33	20.33
			Business rates		
Pine Bluff (Ark.)	41.10	40.91	41.05	41.12	41.13
San Diego (Calif.)	26.54	30.43	30.65	31.10	. . .
Atlanta (Ga.)	53.64	58.82	58.87	58.87	58.81
Chicago (Ill.)	34.12	32.12	31.91	31.91	33.87
Louisville (Ky.)	60.96	61.01	55.87	56.84	55.27
Baltimore (Md.)	43.57	43.57	43.57	43.60	44.97
Boston (Mass.)	43.12	42.78	42.78	42.78	44.10
Grand Rapids (Mich.)	35.29	36.02	35.81	35.88	34.63
Butte (Mont.)	43.82	43.82	43.82	44.07	45.36
Memphis (Tenn.)	54.70	54.70	54.95	54.95	54.95

Source: FCC (1999).

because incumbents face the prospect of having to lease their facilities at rates that do not reflect their sunk costs, the incumbent local carriers' incentives to invest are also reduced substantially.[32] Thus, the ambitious architecture in the 1996 Act for unbundling and mandatory (discounted) resale may have undermined as much competitive activity as it has encouraged. Nonetheless, competition is emerging as the result of the overall policy reforms initiated by the Telecommunications Act.

Figure 2-2. Competitive Local-Exchange Carrier Revenues, 1993–2000

Billions of dollars

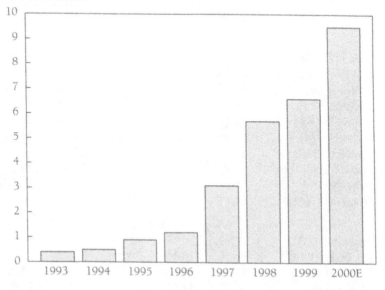

Sources: 1993–98: Federal Communications Commission, *Telephone Trends* (Washington, 2000); 1999–2000E: Credit Suisse/First Boston, Telecom Services, CLECs (New York, June 5, 2000). 2000E is an estimate.

Competitors' Revenues and Market Share. Although the health of competitors can be a misleading guide to the state of competition, in this instance it appears a reasonable starting point. By the revealed preference of consumers, prices adjusted for quality are declining where competitors gain market share from rate regulated incumbent monopolies.[33]

The revenue growth of the competitive local-exchange carriers (CLECs) has accelerated markedly since 1996 according to FCC data.[34] (See figure 2-2.) The small sample size limits the conclusions that may be drawn from these data, but CLEC revenues grew more rapidly after the 1996 Act was passed. In the 1993–95 period, CLEC revenues rose by 50 percent per year

Figure 2-3. Competitive Local-Exchange Carriers, 1997–2000

Percentage of U.S. local exchange market

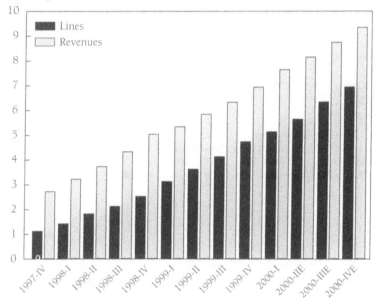

Sources: 1997-IV through 1998-IV, Merrill Lynch, Telecom Services—local (New York, March 11, 1999); 1999-I through 2000-IVE, Credit Suisse/First Boston, Telecom Services, CLECs (New York, June 5, 2000). 2000-IIE through 2000-IVE are estimates.

while in the period just following the act, 1996–97, CLEC revenues grew by 86 percent per year.

The number of local carriers has increased substantially since 1995. One observer has noted that "165 new phone companies [were] spawned by the law."[35] Moreover, the new competitors' combined market share continues to grow strongly, both in access lines and revenues (figure 2-3). The CLECs now appear to have about 9 percent of local revenues and 7 percent of access lines. Although this growth pales in comparison with the pace at which long-distance competition developed, it suggests some progress toward a more competitive marketplace.

Table 2-5. CLEC Equity Returns, 1994–98

Company	March 1999 market cap- italization (millions of dollars)	Initial offering date	Annual growth rate (percent)	S&P 500 adjusted growth rate (percent)
Intermedia	953.70	March-92	21.56	1.25
ICG	908.70	May-92	6.98	–10.89
Winstar	1360.00	Oct.-93	66.37	38.57
GST*	253.70	March-94	2.31	–17.06
S&P 500			20.06	

Source: quote.yahoo.com.
Note: *GST returns calculated from IPO date in March 1994. $40,000 invested in 4 CLECs returns $163,547 in five years. $40,000 invested in S&P 500 returns $99,781.88. Abnormal CLEC portfolio return of 10.39 percent annually.

CLEC Stock Market Performance. The modest market shares of the CLECs reflect the slow start of local competition in the United States both because of legal battles over implementation of the act and the substantial capital costs and time required to build local networks. The equity-market performance of the small number of publicly listed CLECs during the five-year period 1994–98 provides at least modestly positive evidence of the financial effects of the 1996 Act on competitive local-exchange carriers (table 2-5). Only four companies can be charted throughout this period,[36] a span during which the act was drafted, debated, amended, passed by Congress, signed by the president, enacted by the FCC, and litigated in federal courts.

Since the act ostensibly aimed to enhance competition in the local-exchange market, it is reasonable to expect that firms specializing in providing such service would enjoy windfall gains during this period. However, while all four of the listed companies produced positive returns for shareholders, only two (Winstar and Intermedia) outperformed the S&P500 Index, which grew at an average rate of 20.06 percent per year. Winstar's performance was sufficiently in excess of the market return as to make the performance of the portfolio of CLEC stocks superior to the market as a whole. Had one invested $10,000 in each of the CLECs at the beginning of 1994, the equally weighted portfolio would have been worth $179,226 at the end of 1998. The same amount ($40,000) invested in the S&P500 would have grown to just under $100,000. Hence, capital gains in the small, publicly listed CLEC sector were more than twice that for the S&P 500.

Some of this supranormal return is likely a risk premium for holding CLEC stocks, all of which have betas in excess of one.[37] Still, CLEC returns appear to be in excess of the market as a whole even with this adjustment. It is also of interest that the star performer in this CLEC group was Winstar, a wireless firm able to offer facilities-based competition as opposed to service over leased portions of incumbents' networks.

Further information can be gleaned from the stock market evidence on CLECs shown in table 2-6. Even though there is only a small sample of CLECs publicly listed throughout the relevant period, the sample becomes substantially larger over time. By March 2000, listed CLECs had a market capitalization exceeding $130 billion, up from the sector total of under $3 billion in 1996. By this measure competitive entry into local telecommunications has been impressive. By way of comparison, throughout the years following the 1984 Cable Communications Policy Act (legislation promising greater competition in local cable markets), no new public firm—of any size—emerged to offer head-to-head competition in cable service. Unfortunately, since March 2000, the value of these CLEC equities has plunged, and many of the new companies have encountered serious financial difficulties.

Perhaps better indication of the progress of the new competitive local carriers is their willingness to devote real capital resources to local telecommunications markets. Table 2-7 provides the available data on capital spending by all listed CLECs from 1996 to 1999. These companies spent at a rate of about $14 billion per year during that period, or more than the recent levels for all U.S. commercial mobile wireless carriers.[38] If these

Table 2-6. Market Capitalization for U.S. Competitive Local-Exchange Carriers: March 22, 2000

Company	Market capitalization (millions of dollars)	Initial offering date
Intermedia Communications	2,889	March-92
ICG Communications	1,552	May-92
Winstar Communications, Inc.	4,786	Oct.-93
GST Telecommunications	334.7	March-94
e.spire Communications, Inc.	619.9	March-95
CTC Communications Group	1,030	Aug.-95
McLeodUSA, Inc.	13,043	June-96
Advanced Radio Telecom	958.5	Nov.-96
RCN Corporation	4,818	Sept.-97
ITC/DeltaCom, Inc.	2,261	Oct.-97
Nextlink Communications	11,200	Oct.-97
Electric Lightwave, Inc.	1,214	Nov.-97
Teligent, Inc.	4,735	Nov.-97
Worldpages.com	193.3	Feb.-98
Level 3 Communications, Inc.	38,115	April-98
US LEC Corp.	1,144	April-98
Adelphia Business Solutions/Hyperion	4,080	May-98
MGC Communications	1,480	May-98
Allegiance Telecom	8,237	July-98
Caprock Telecommunications	1,580	Aug.-98
12 others	28,885.8	Jan. 99–March 00
Total market cap	133,156.2	

Sources: quote.yahoo.com and www.clec.com.

CLECs are able to complete the roll-out of their networks, many of them should be much more potent competitors. Such levels of capital spending also call into question the need for whole-sale unbundling of incumbents' networks.

Local Competition in Canada. At this juncture, it is difficult to assess the impact of the Canadian policy toward local competition because the CRTC only began to allow competition in May

Table 2-7. Capital Spending by Competitive Local-Exchange Carriers, 1996–99

Millions of dollars

Company	1996	1997	1998	1999
Adelphia Business Solutions	77	121	349	478
Allegiance Telecom, Inc.		22	114	273
Advanced Radio Telecom	17	17	11	NA
US LEC Corp.		13	47	57
Concentric Networks		5	23	42
Convergent Communications, Inc.		2	20	49
Covad Communications		2	60	208
CTC Communications Group	1	6	36	NA
CapRock Telecommunications	10	14	36	201
Electric Lightwave, Inc.	56	104	201	180
e.spire Communications, Inc.	108	135	249	287
Focal Communications Corp.		12	64	129
GST Telecommunications	98	214	248	271
ICG Telecommunications Inc.	175	287	395	765
Intermedia Communications	147	260	473	680
IDT Corp.	12	NA	72	95
ITC DeltaCom Inc.	6	44	148	166
Level 3 Communications, Inc.		26	910	3,311
McLeod USA Inc.	174	601	340	1,317
Metromedia Fiber Network	107	19	115	549
MGC Communications	4	58	82	83
Network Access Solutions			5	55
Network Plus CP	2	3	11	94
NorthPoint Communications		1	42	197
Nextlink Communications	78	254	594	1,127
Primus		40	76	111
RCN Corp.		79	286	526
RSL		36	182	208
Rhythms NetConnections		1	10	193
SpeedUS.Com			31	20
Teligent, Inc.	·	10	183	262
Time Warner Telecom	145	127	126	221
Winstar Communications, Inc.	46	220	402	1,278
Other	12	29	99	190
Total	1,275	2,762	6,040	13,623

Source: Company financial statements from quote.yahoo.com.

1997.[39] Moreover, given much slower economic recovery from the 1990–91 recession, Canada's cable television companies have only recently evidenced an interest in offering telecommunications services. Thus, it is too early to observe the result of Canada's much less intrusive policy for liberalizing local telecommunications markets.

The Wireless Sector—Competition without Regulation

It may be instructive to examine parallel developments in a similar telecommunications sector where entry by facilities-based operators has been the sole driver of increasing competition. This entry, moreover, has occurred without the elaborate regulatory protections afforded new competitors in local-exchange markets. We refer to wireless telephony.

Competition in U.S. commercial wireless services had been slow to develop until very recently. For more than a decade, the United States had but two wireless (cellular) providers in each market because the FCC decision allocated only two 25-MHz licenses to each market. In addition, unlike Europe or Canada, the United States licensed wireless services on a geographically fragmented basis. In 1990, however, the FCC began to allocate microwave spectrum for personal communications service (PCS), a cellular substitute.[40] In 1995 licenses were auctioned to assign PCS licenses in the A and B bands (each allocated 30 MHz). More recent wireless telephone auctions have assigned licenses in the C (30 MHz), D (10 MHz), E (10MHz), and F (10 MHz) bands. Furthermore, an entrepreneurial company—Nextel—succeeded in obtaining FCC approval to reallocate approximately 10 MHz of spectrum from a dispatch-mode service to enhanced specialized mobile radio (ESMR), a commercial wireless service competitive with cellular and PCS (table 2-8). Hence, nine licenses are available in local wireless telephone markets, markets essentially unregulated with respect to pricing. Licenses may be aggregated, but given the FCC's 45 MHz "spectrum cap," four or more competitors are maintained per market.

Table 2-8. U.S. Wireless Telephone Licenses, 2000

Service	License	Bandwidth	Number of license areas
Cellular	A	25 MHz	734
Cellular	B	25 MHz	734
PCS	A	30 MHz	51
PCS	B	30 MHz	51
PCS	C	30 MHz	493
PCS	D	10 MHz	493
PCS	E	10 MHz	493
PCS	F	10 MHz	493
ESMR	—	10 MHz	—

Source: FCC (1996).

Wireless Rates. The first of the new PCS services began in late 1995. Since that time, a large number of companies have begun building facilities. All but twenty-eight of the top one hundred metropolitan markets in the United States had at least five wireless competitors in early 1999—two cellular providers, two to four PCS services, and Nextel.[41] The effect of the resulting competition on wireless rates in the United States has been stunning (figure 2-4). Throughout the 1984–95 period, real, inflation-adjusted cellular rates fell at a rate of 3 to 4 percent per year.[42] Between 1995 and 1999, however, real cellular rates fell at a rate of 17 percent per year as PCS service providers offered service at per minute prices less than 50 percent of prevailing cellular rates. There is evidence that with open entry just one new player is sufficient to drive rates sharply lower.[43]

The only new regulation imposed on the wireless industry by the 1996 Telecommunications Act was the requirement for "reciprocal compensation" on wireless-wireline interconnection. Before 1996 the rates for exchanging traffic were often in excess of 2 cents per minute; today, they are in the range of 0.5 to 0.7 cent per minute because wireless companies are afforded the

Figure 2-4. Real Wireless Prices, 1993–99

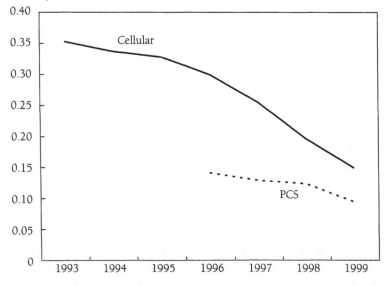

Dollars per minute (1982–84 $)

Source: Leibowitz (2000).

same interconnection rates as the new CLECs. Otherwise, competition has increased dramatically without unbundling, resale, or other new forms of regulation. Indeed, wireless services are essentially exempt from retail-price regulation.

Subscriber Penetration. The United States has lagged behind some European countries in cellular penetration, in large part because of its fragmented approach to licensing and the resulting roaming charges as well as FCC-mandated charges on both calling and receiving parties. The development of national service by AT&T, Sprint, Nextel, Bell Atlantic-Vodafone, SBC, Voicestream, and U.S. Cellular, however, has recently led to national one-rate pricing. While subscriber penetration is only

40 percent of the population, growth has accelerated with price declines resulting from entry of PCS operators.

As long as the United States keeps residential wireline rates low and does not allow local carriers to charge for local usage, the equilibrium U.S. penetration may remain below that realized by Hong Kong, Italy, and the Scandinavian countries. However, we have not yet witnessed the full effect of the recent sharp decline in U.S. wireless rates. Nor has the United States been able to implement "calling-party-pays" tariffs, motivating wireless subscribers to leave their lithium-ion-powered handsets on continuously to receive calls.

U.S regulators have provided us with an illuminating experiment. In wireless telephony, additional facilities-based competitors were licensed and then permitted to operate without rate controls, unbundling requirements, or mandated resale. Local wireline telephony, however, was approached as a natural monopoly market that required extensive regulation to jump-start competitive entry. The rules establishing cost-based access to the incumbents' facilities proved difficult to devise and arduous to implement, particularly given the litigious nature of regulated industries in the United States. The simpler approach to regulation not only has administrative efficiencies to recommend it but also appears to generate greater competition. Facilities-based competition may develop in wireless telephony rather than in wired systems because the latter must build connections to the final subscriber. But both sets of service providers now co-exist, pitting wireless against wireline carriers in a head-to-head rivalry for local dial-tone service, thereby eroding the case for regulatory supervision of local competition.[44]

The Political Consensus behind the U.S. Telecommunications Act

The Telecommunications Act was the product of realpolitik. Reform involves compromises and political bargains such that an actual majority—or supermajority to circumnavigate the veto power of various interest groups and committee chairs—can form a coalition to enact law. In political institutions interest groups jockey for advantage, angling for better deals. Often the interests of incumbent officeholders in continuing contentious legislation into future legislative sessions (where support groups can be cajoled or threatened, and electoral benefits thereby extracted), combines with the interests of reform opponents to block legislation altogether. The status quo is rarely without a considerable number of friends—which is how it *became* the status quo. For this reason, countless efforts to "update" the 1934 Communications Act had been repulsed, including the ambitious effort by Representative Lionel van Deerlin (D-Calif.), then chair of the House Subcommittee on Finance and Telecommunications, to chart a new course in the 1976–80 period.[45]

To overcome such natural inertia in controversial legislation, it is helpful to have a powerful countervailing motivation for Congress to act. A public emergency is the classic legislation-moving pressure, a situation when the standard reasons for not legislating are momentarily overwhelmed by political actors who seize the opportunity (partly out of desire to seize credit for forging a solution, partly out of fear of appearing unresponsive or "out of touch"). Yet, no great crisis gripped the public in 1996; telecommunications reform was no more visible than in previous years. Why did legislation pass at this moment and not before?

Although motivation factors are sometimes difficult to pinpoint, an important moment of decision was at hand in telecommunications law: any policy shift governing the AT&T decree's line-of-business restrictions would have to come just about the time the telecommunications act was passed. Otherwise, the federal courts would likely soon relax the restrictions. Congress pre-empted this in the act, moving jurisdiction away from Judge Harold Greene, and seizing it for Congress and an agency it oversees, the FCC.

The RBOCs had been constrained by the consent decree that ended the AT&T antitrust suit in 1982. The restrictions were challenged almost at once by the RBOCs, however, and the challenge gained momentum following a 1987 Department of Justice report documenting that the telecommunications market was changing so rapidly that the rationale for restrictions was becoming dubious.[46] Between that time and the mid-1990s there was intense legal skirmishing. As described by Peter Huber, author of the 1987 report:

> [Judge Greene's] courtroom operated as a shadow FCC, an independent authority that scrutinized, cajoled, hectored, and prosecuted. There were hundreds of motions, complaints, and other requests to enforce, modify or interpret. The Justice Department issued thousands of advisory letters. The court received over six thousand briefs. Thirteen groups of consolidated appeals were carried to a federal appellate court in Washington. The Supreme Court received half a dozen divestiture-related petitions for review. . . .
>
> A 1995 Justice Department proposal to grant limited relief to two local phone companies in Chicago and Grand Rapids occupied twice as much paper as the entire consent decree that broke up the national Bell System. This Son-of-Sam decree addressed network information, billing services, and customer lists. It devoted four paragraphs to regulations for marketing services to business customers and another three to marketing to residential customers. The Justice Department itself was to review and approve a written script used by Ameritech to sell interexchange service. Two paragraphs were required to spell out how Ameritech would list local competitors in its white pages.
>
> The 1996 Telecommunications Act put an end to all this. It transferred authority over the key line-of-business restrictions to the FCC, and it established a process and timetable for getting rid of them all.[47]

Though certain of the RBOC requests were granted by the courts (typically the D.C. Circuit overturning Judge Greene, who viewed the RBOC filings with great skepticism), the activity generated by interest groups fighting for position created a demand for judicial rulings rather than regulatory—or legislative—favors. This state of affairs was undoubtedly less than optimal in

**Figure 2-5. Soft Money and PAC Contributions
in 1996 and 1998 Election Cycles**

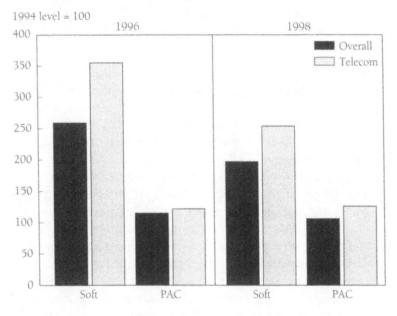

Source: Data from Center for Responsive Politics, Washington, 1999.

the opinion of incumbent members of Congress. In short, passing the Telecommunications Act, moving the marketplace back
to *Congress's* line-of-business restrictions, was a popular, bipartisan objective among legislators. This propelled legislation that
had been stalled for, literally, decades. We may now judge whether
Congress's self-interested objective has been met.

The evidence suggests that it has. In the 1996 and 1998 election cycles, federal political contributions by telecommunications firms rose absolutely and relative to the overall rise in
political giving, according to data supplied by the Center for
Responsive Politics (figure 2-5). Indeed, in both cycles, both
categories (soft money and PAC donations) of telecommunications spending increased. This 4 for 4 outcome could be achieved

by random chance just 6 times in 100. This increase in political contributions is all the more impressive in light of the act's announced goal of "de-regulation." Rent seeking predictably subsides as government intervention recedes. In this instance, reported expenditures on political contributions actually increased.

Add to this quantifiable political gain the fact that the Telecommunications Act has provided a platform for an exceptionally newsworthy set of public issues, from the major competitive issues discussed in this chapter to the "hot button" social issues in the act, from TV violence (and the V-chip) to Internet indecency. Even failings attributed (rightly or wrongly) to the act (say, cable rate increases) have afforded the opportunity for high-profile hearings and voluminous incumbent publicity. It is not an overstatement to label the Telecommunications Act close to an unmitigated political success for Congress as a whole.[48]

Conclusions

A sober assessment of the effects of the 1996 U.S. Telecommunications Act reveals that the legislation has promoted entry into local telecommunications, though at a rather slow rate. It has not, however, yet allowed the RBOCs to enter most long-distance markets and has thereby slowed the development of interexchange competition. By contrast, Canada has not attempted to divide its telephone industry into a set of local carriers and a set of long-distance companies, yet it has achieved as much long-distance competition in seven years as the United States has achieved during a quarter century.

We conclude that the 1996 U.S. Act, while flawed, still scores relatively well in comparison with previous U.S. telecommunications laws. Increasingly, customers are facing choices among service suppliers. Where head-to-head rivalry has developed, rates are falling and choice increases. Entry by new rivals seems to be speeding deployment of enhanced services, including high-speed broadband access for business and residential customers.

Capital markets—always looking to the future—indicate that competitive forces in the United States will intensify. Billions of dollars are now being employed on bets that firms offering competitive local telephone and cable service will prosper and grow. Regardless of the impatience justifiably exhibited with monopoly services in the interim, it must be pointed out that previous legislation—two comprehensive and much heralded Cable Acts in 1984 and 1992, for instance—never succeeded in producing the service rivalry that now exists on the competitive fringe of local telephone markets.

It is still too early to provide a definite measure of the importance of the 1996 U.S. Act or the 1997 Canadian local competition decision. Unfortunately, market forces are not labeled, *Made in the Act*. But having observed that previous "reforms" have produced demonstrably counterproductive impacts for consumers, it is telling that such problems have not yet arisen in the wake of the 1996 U.S. legislation. Indeed, the failings of the Telecommunications Act in promoting competition are likely to be found in its conservatism. The measure did not liberalize radio spectrum allocation nor move aggressively to promote long-distance entry by the RBOCs. It mandated extensive "safeguards" and led the FCC to micromanage reforms so tightly that the leading U.S. regulatory economist, Alfred Kahn, has proposed "deregulating the process of deregulation."[49]

3

The Liberalization of European Telecommunications

Martin Cave
Luigi Prosperetti

I n this chapter we provide an overview of the liberalization
process in European telecommunications and point out what
we consider the major items on its agenda for the next few
years. Given the comparative nature of this study, we also point
out some differences from and common features with the U.S.
experience.

Drawing such parallels is of obvious analytical and practical
interest. It also provides some curious insights, as North Americans and Europeans appear to be concerned with opposite sides
of the same coin: to dissolve or not to dissolve the Federal Communications Commission (FCC)? To create or not to create a
European Regulatory Authority? We also see a tendency for both
sides to find the grass greener on the other side of the Atlantic
(for example, on tariff rebalancing).

But we consider chiefly the European experience. By and large,
we agree with the point made by Romano Prodi, president of the

Paper presented to the CEPS and Joint Center for Regulatory Studies workshop on
Transatlantic Regulatory Issues, 25–26 April, 2000, Washington, D.C. We grate-
fully acknowledge research assistance by Matteo Merini and Emma Putzu.

European Commission, at the March 2000 European Union (EU) summit in Lisbon, which was centered on information society issues: telecommunications liberalization in Europe is a success story. The data quoted by Prodi bear this judgment out quite well: from 1998 (the year in which European telecommunications entered a "full competition" regime) to 1999, international call prices fell by an average of 40 percent ; long distance by 30 percent; regional prices by 13 percent. There are now 460 telecom operators in Europe. Between 1998 and 2000 the total telecom services market grew an estimated 12.6 percent, to 161 billion euros.[1]

However, the European record is more mixed than these data suggest. Competition has not led to the widespread deployment of alternative infrastructures, and this outcome has kept leased-line prices at a very high level. This, in turn, has hampered the growth of Internet penetration, which is increasing but still lags substantially behind that in the United States. In mobile communications, however, Europe seems to have gotten most of the regulatory choices right and now enjoys a lead in penetration.

In this chapter we outline the EU telecommunications regulatory package in the light of some basic institutional factors. We then analyze the implementation of the package in fixed-line communications, tackling licensing, interconnection, local loop unbundling, universal service, and leased lines. We discuss the implementation of the package in mobile communications and show that a *laissez faire* approach worked quite well in the past but now is in the course of a somewhat inevitable reversal. We also summarize the outcome of the 1999 Communications Policy Review initiated in November 1999 by the EU Commission. Before concluding, we examine what we see as the two principal issues on the agenda: enforcement, with the associated institutional problems; and the balance between ex ante rules and ex post antitrust action.

The EU Telecommunications Regulatory Package

Telecommunications reform in Europe has not been easy to accomplish. Two features that are crucial to understand if one seeks to achieve a balanced view of this process are the complex decision structure of the EU and the extent of public ownership across the Continent.

The European Legislative Process. Europe has a multilayered institutional framework, which makes for complex law and rulemaking processes. At its core sits the European Commission, a nonelected body whose term runs for five years. Only the Commission has the power to initiate EU legislation. Such legislation must, however, be concerned only with issues that involve, or may involve, more than one member state and must aim to achieve an objective set out in the EU Treaty. The "interstate" element may arise because actual or potential trade issues are involved, or because there are spillover effects across countries arising from an activity being conducted within a single member state. Other matters are dealt with at a national level: this is the essence of many discussions about "subsidiarity" in Europe.

The most common procedure when the Commission seeks to legislate is as follows: it prepares a draft and sends it to the European Parliament, an elected body, which provides its observations. The Commission then prepares a revised draft, and there is another loop involving the European Parliament. Provided such a process turns out to be convergent, legislation—usually in the form of a directive—is eventually enacted by the Parliament and European Council of Ministers (that is, the meeting of all the ministers handling the issue at hand in member states' governments). The whole process is a lengthy one, usually taking between eighteen months and two years, and it may be much longer on highly contentious issues. Such directives, which are then the joint work of the Commission, Parliament, and the Council of Ministers, are often referred to as harmonization directives and are based on article 96 of the European Treaty.

Once a directive comes into force at the EU level, it is transposed according to each country's legal system by national legislation passed by the member state's parliament: this may diverge, sometimes substantially, from the directive it is supposed to enact. However, a directive takes immediate effect within member countries upon its publication in the *EU Official Journal* and can be referred to by national courts. This sort of "double track" for the implementation of European directives in national legal systems gives several grounds for delay and extended legal proceedings. The Commission has the power to open infringement proceedings against member states, but these actions take a very long time, and the sanctions lack bite.

Under a separate legal procedure the Commission can also enact a directive on its own when it deems it necessary to safeguard the effectiveness of the competition provisions of the European Treaty, for example, to avoid possible abuses of dominant position or to remove special or exclusive rights bestowed by national legislation on an operator. Such directives are known as liberalization directives and are based on article 86 of the treaty.

The Commission has had recourse to both sets of instruments to liberalize telecommunications: typically, after a very long gestation, a framework directive is produced on the basis of article 96. After some time, the Commission "discovers" that some national legislation is preventing the directive from achieving its objectives and steps in with an article 86 directive, swiftly removing such obstacles.

Considering EU legislation in telecommunications in perspective, we see that many significant steps were accomplished through the Commission's own directives: this was the case with the 1990 services directive, which liberalized supply across the EU of all services except voice telephony, and of the 1996 competition directive, which liberalized the rest as of January 1, 1998.

Public Ownership and the Liberalization Process. The complexities in opening the European telecommunications markets can be better appreciated if we consider the extensive role that state ownership had (and indeed has) in European telecommunications. In 1993, when the liberalization process had already run for about six years, only in Britain and Spain was the share of equity controlled by the state below 40 percent. Even in 1999, the majority of shares of certain very large operators such as

France Télécom and Deutsche Telekom were still controlled by the state. This was also true of Austria, Finland, Sweden, and Belgium.

The 1998 Liberalization Package. It is useful to assess against this background the 1998 legislative package, which aimed at liberalizing telecommunications in the European Union.[2] For ten years or more before that, a series of green papers, directives, recommendations, and other interventions imposed obligations on member states concerning equipment markets, regulatory structures, value-added services, and the regulation of infrastructure and service competition where it existed. But in 1998 the obligation was imposed on governments to liberalize entry into their telecommunications markets (except for those few for which extensions were granted). In our view, the edifice of regulation then enacted represented a deliberate framework of structural and behavioral regulation appropriate to the development of the European industry at the time. It divided responsibilities between the Commission and national regulatory authorities (NRAs) in a way designed to give early adopters of new competitive arrangements the opportunity to press ahead, while giving the Commission powers to force laggards to meet their obligations. General European competition law—articles 81 and 82 of the treaty prohibiting cartels and abuse of a dominant position—coexisted with industry-specific measures.

However adequate the 1998 framework was at a conceptual level, serious defects appeared in its implementation. Liberalization and harmonization directives first have to be transposed into national legislation to take effect in the member states. The transposition process took nearly two years, but in October 1999 the Commission was able to report that it was largely complete. As noted above, the directives give member states considerable latitude in implementation: for example, the licensing directive permits ample variation on the requirements imposed on new entrants, and despite a requirement in the interconnection directive that interconnection charges be cost based, interconnection charges within the EU vary greatly.

These problems of implementation have permitted the emergence of a substantial gap between theory and practice.[3] To illustrate this proposition, we first set out in broad terms what we understand to be the strategy underlying the directives and then examine in more detail how it has been realized and its effects.

Achieving the goals of liberalizing the industry against the (initial) wishes of most incumbent operators and of many member states has required substantial regulatory intervention from the Commission. Moreover, the regulatory framework had to be flexible enough to cover member states proceeding at quite different rates. In the early stages of liberalization—the transition to competition—it is necessary to constrain the former monopolists considerably. Gradually this restraint can give way to a state of normalization, as competition is established and regulatory intervention diminishes.

The key structural requirement is that, in markets in which there are no technical restrictions on entry, arising, for example, from spectrum scarcity, licenses should be granted to operators subject to the minimum of conditions.

For behavioral regulation, three behavioral instruments are required in the early stages of liberalization:

—*Control of retail prices.* This is necessary when the dominant firm exercises market power at the retail level; in the absence of retail price controls, customers will be significantly disadvantaged. Member states have historically fulfilled this consumer protection function, though under monopoly conditions, the controlled tariffs were seriously unbalanced with respect to cost. However, as competitive supply emerges at the retail level, possibly from firms relying largely on infrastructure belonging to others, the necessity for retail price controls in effectively competitive markets may disappear.

—*Control of access prices.* In order to keep all subscribers connected with each other in the presence of competing networks, operators require *access* to one another's networks to complete their customers' calls. This requires a system of interoperator wholesale or network access prices. Especially in the early stages of competition, entrants will require significant access to the dominant incumbent's network, and this relationship will almost inevitably necessitate regulatory intervention. However, as infrastructure is duplicated (initially the infrastructure necessary for long-distance and international conveyance), the need for direct price regulation of certain network facilities diminishes. Interconnection has been central to the development of competition within the EU, and the Commission has been heavily involved.

—*Universal service obligation.* Governments have typically imposed a universal service obligation (USO) requiring the historic telecommunications operator to provide service to all parts of the country at a uniform price, despite the presence of significant cost differences. Firms entering the market without such an obligation have a strong incentive to focus on low-cost, "profitable" customers, putting the USO operator at a disadvantage. Pressure may therefore build up to equalize the situation, perhaps by calculating the net cost of the USO borne by the dominant operator and then sharing the cost among all operators. For universal service, the concern has been, first that it would be used as a pretext for delaying competition, and second that high USO contributions imposed on entrants would choke off competitors.

We have noted that regulatory policy in relation to the control of retail and wholesale prices and the allocation of universal service obligations is likely to change throughout the three phases of liberalization: monopoly (or duopoly), transition, and normalization. Table 3-1 sets out how each of the three regulatory instruments might develop.

Implementation of the Package in Relation to Fixed-Line Services

We now examine in greater detail the extent to which the reality has matched the conceptual framework.

Licensing. Under the licensing directive, licensing regimes should be light touch and transparent. These requirements imply that

Table 3-1. Stages of Regulation

Item	Monopoly/duopoly	Transition	Normalization
Retail price control	Controls on all services	Gradual relaxation of controls	No controls
Access pricing	Not relevant, or arbitrary pricing of small range of services	Services gradually decontrolled; unbundling; use of price caps	Only "bottleneck" services, such as call termination, controlled
Universal service obligations	Borne by incumbent	Costed and shared (or ignored if not material)	As transition phase, with possibility of competitive provision

general authorizations should be favored over individual licenses, that potential licensees should know what conditions have to be satisfied, and that onerous conditions should be avoided as far as possible; in addition, licenses should be issued quickly and fees kept to a minimum.

In the 1999 Implementation Report, published in October 1999, the Commission noted considerable divergences between national licensing regimes "ranging from light as possible, where operators are free to enter the market without formality . . . or are required simply to register . . . or notify . . . their intention to do so . . . , to the extremely heavy, where individual licences are the rule and in some cases a government minister is required to sign every licence."[4] (This conclusion rested on a comparative analysis of national licensing regimes summarized by Laura Pontiggia and Ann Vandenbroucke, which focused upon two aspects : first, the level of segmentation of authorizations and the extent to which individual licenses are required and, second, the volume of information required about applicants.)[5]

Some member states have a regime in which an authorization permits a licensee to provide all telecommunications networks

Table 3-2. Licensing Variations in the European Union

Use of individual licenses	Level of segmentation		
	Low	Medium	High
High		Belgium Germany Luxembourg Spain	Italy United Kingdom
Medium	Ireland		France
None/low	Denmark Finland Norway Sweden Netherlands		

Source: Pontiggia and Vandenbroucke (2000).

and services, excluding services that use scarce frequency resources and therefore require individual licenses. This low level of segmentation can be combined with licensing under general authorizations rather than individual licenses. At the opposite extreme, other member states stipulate as many as twelve categories of authorization, each with its own licensing conditions and procedures, and rely heavily on the issuing of individual licenses that in some cases are tailor made to fit geographical service areas and other characteristics. Table 3-2 illustrates the diversity of practice.

Similar differences exist concerning information requested before market entry. The study found that member states with high levels of segmentation also tended to impose high informational requirements. For example, two member states require information about the operator's marketing plan, three require a financial plan or investment plan, three require a plan of recruitment, and nine a schedule for roll-out.

The Implementation Report shows that procedures for licensing go beyond the limits laid down in the licensing directive in three countries, and that in Italy, licensing procedures appeared to be protracted, heavy, and lacking in transparency. The Commission also concluded that in three countries fees and charges appeared to be higher than the administrative costs incurred.

The remedy for such problems is fairly straightforward—the achievement of harmonization by adopting a common set of requirements based on the procedures of member states operating a light-touch regime, or the introduction of a one-stop shopping regime, in which a license granted in one member state offers equivalent entitlements throughout the whole of the EU. Such principles are indeed reaffirmed in the 1999 Communications Review by the European Commission, which will be discussed below.

Despite this uneven performance, licensing practices have not, in general, acted as an obstacle to market entry in fixed-line telecommunications. Thus in Italy, where both market segmentation and the use of individual licenses is high, where procedures are lacking in transparency, and information requirements are exacting, about one hundred fixed-line licenses have been issued, and entry is abundant.

A different, but highly complementary, legal issue concerns the concession of rights of way. Given the high levels of urban congestion across Europe, these rights often turn out to be entry barriers of large practical relevance. The legal structure of the problem is complex in some countries, since it touches on the constitutional rights of local authorities. According to the Commission, in France, Spain, Italy, and Belgium considerable delays occur in granting public or private rights of way.[6]

Interconnection. Access to networks is of fundamental importance for the development of effective competition in the EU telecommunications industry. All infrastructure operators experience growing demands from entrants to interconnect with them.

An important component of a policy to promote effective competition is a regulatory environment guaranteeing that competitors have access to networks that they cannot duplicate. Fair access to such facilities, and in particular fair access prices, will generally improve economic efficiency by easing competition in markets upstream and downstream of the bottleneck. This is true whether the industry is vertically separated or not. At a more general level, access pricing is part of the antitrust concern with market foreclosure, which is central in the so-called essential facilities doctrine. This emerged in the United States and has spread (using different terminology) to European courts. However, sector-specific legislation in Europe typically requires all operators, including incumbents, to open their networks to competition.

The price that entrants have to pay for access to the incumbent's network is crucial to their commercial success, yet the incumbent has two motives for charging high access prices. The first is a simple desire to maximize monopoly profits; the second, which arises if a network owner is also competing in the retail market, is the desire to raise its rivals' costs and maintain a dominant position in that market.

These considerations have meant that regulators have to intervene in access pricing by imposing detailed prices for the use of the elements of the network or by limiting the overall revenues that the network owner can collect to those that are necessary for the recovery of its costs. Either approach involves a detailed analysis by the regulator of the costs incurred by the network. The first best solution would involve setting access prices on the basis of marginal cost. This may need to be supplemented by a markup if such pricing would prevent the incumbent from breaking even.

In Europe there is the obvious danger that new entrants will choose to enter the profitable long-distance market even though they are less efficient at providing long-distance conveyance than the incumbent. This will also deny to the incumbent the call revenues that are necessary to cross-subsidize access, where tariffs are unbalanced. One way of eliminating this possibility is by allowing the incumbent to charge new entrants for access to its network at rates that take account of the lack of balance in the tariff structure. Thus the incumbent could be allowed to make a charge for call termination that covers not only the costs of that call termination but also includes an additional element comprising the excess profit that the incumbent *would have made* by providing the call at retail prices, and which it needs in order to cross-subsidize other services such as line rentals. This approach to access pricing is known as the efficient component pricing rule, or ECPR.

If an interconnection charging regime of this kind were introduced, then an entrant would be able to gain a profitable foothold in the industry only if its costs in supplying the services that it provides itself, rather than buys from the incumbent, are less than those of the incumbent. Such a rule would, in consequence, encourage entry only where the entrant is more efficient than the incumbent.

The regulation of interconnection pricing in the EU has not gone down this path. The interconnection directive requires that charges for interconnection follow the principles of transparency and cost orientation. The first principle implies the publication of a reference interconnection offer. As a corollary, operators with significant market power are required to keep separate accounts for their wholesale or network activity and for other activities, including retailing.

Cost orientation turned out, however, to be an excessively vague phrase, permitting excessive interconnection charges. There are two major reasons for this situation. First, the interconnection directive took a rather catholic view of cost standards, citing "fully distributed costs, long-run incremental costs (LRIC), marginal costs, stand-alone costs, embedded direct costs." Each of those can be measured, according to the directive,[7] on the basis of a historic or forward-looking cost basis. This was obviously unavoidable, as accounting standards differ rather widely across the EU (for example, Italy relies on historic costs, while the United Kingdom has a current cost accounting system), and their differences are firmly rooted in national tax codes. Second, analyzing cost data is a highly complex business for a regulator, and regulatory bodies in member countries—with the obvious exception of the U.K. Office of Telecommunications (OFTEL)—are mostly less than five years old. Some of them suffer also from scarcity of resources.

So cost orientation in many cases turned out to be a general philosophy rather than a practical approach. Until cost data are available based on the Commission's preferred methodology—long-run incremental cost (LRIC)—the Commission publishes recommended "best current practice" interconnection charges, based on the average of the member states with the lowest charges. Actual values reported by the Commission for local and double transit interconnection show rates that are declining but exhibit considerable variation across member states.[8]

Double transit interconnection charges (the ones that really matter in most EU countries, where competitive local access providers are developing slowly) dropped rather fast for the smaller countries, which in 1997 had charges much higher than the benchmark values—notably in Belgium and Finland. The record of larger countries is more mixed: a substantial decrease took place in several countries. In early 2000 charges were still much higher than the benchmark in all countries, except for the United Kingdom, Sweden, Luxembourg, and the Netherlands. All in all, the benchmarking approach has been beneficial, but the largest decreases take place at the outset of its application: unsurprisingly, the device seems to be exhibiting a sort of regression toward the mean.

The main issue is whether the system employed is serviceable. It relies on "peer competition" among regulators to bring down the charges. It is clearly not in line with the cost-based approach embodied in the interconnection directive as the basis for the regulation of interconnection in Europe. From a theoretical point of view, and similarly from a legal point of view, its efficiency might be questioned, but, at least for a while, it seems to have had results.

Local Loop Unbundling. The 1999 Implementation Report highlights a lack of competition in local access markets in all member states, even though 375 operators in the EU offered local network services. There are several possible reasons:

—The unbalanced nature of tariffs in many member states, leading to a reluctance to invest, especially when carrier preselection may deprive the local operator of "ownership" of the subscriber's long-distance and international calls;

—The failure of fixed radio to provide a viable access alternative;

—The delay in the provision of competition for cable TV operators in member states that are already highly cabled (as a result of the significant incremental investments required), as well as in member states where cable networks are being rolled out.

This has encouraged NRAs to look to local loop unbundling as a means of introducing competition into the market for access, especially the market for high-speed Internet access for both residential subscribers and small and medium-size enterprises. The Commission's 1999 Communications Review indicated that this was an item on the agenda, but opinion hardened around the view that more urgent action was required.[9]

Table 3-3 describes an interesting variety of pricing mechanisms proposed: thus in Austria, Germany, Sweden, and the United Kingdom prices for an unbundled loop will be set on the basis of forward-looking costs. Italy sided with fully allocated historic costs, with the rather peculiar motivation that otherwise new entrants would not have enjoyed the same economies of scale of the incumbent. In Denmark, by contrast, "retail minus" pricing is employed. The Netherlands has adopted a more nuanced regime in which the price of the loop increases over a five-year period, reflecting the regulator's desire increasingly to encourage competing operators to "make" rather than "buy."

Rather than wait until the implementation of the new regulatory package in 2003, the Commission gave early consideration in a working document to proposals for a recommendation to mandate local loop unbundling. It identified three technologies: unbundled access to the copper pair; shared use of the copper line; and high-speed bit-stream.

For contingent historical reasons, each was subject to different regulations under the current package. It is clear that the interconnection directive does not apply to full unbundling of the local loop. In the case of shared use of the copper line, the voice telephony and the interconnection directives provide the

Table 3-3. Unbundled Local Loops in EU Member States[a]

Country	Status[b]	Basis for price
Austria	12.4€/month	Price based on current valuation of assets
Belgium	Consultation	. . .
Denmark	8.23€/month	Price based on telephone line rental
Finland	5–25€/month	Price based on current valuation of assets
France	Under consideration	. . .
Germany	13€/month	Price set by regulator based on forward-looking long-run incremental cost
Greece	. . .	. . .
Ireland	Consultation	. . .
Italy	Proposed by 2000	Price based on historic costs
Luxembourg	. . .	. . .
Netherlands	Less than 15.4€/month	Phased pricing set by OPTA moving from historic costs to current costs in 5 years
Portugal	. . .	. . .
Spain	Line sharing access can be negotiated	. . .
Sweden	Proposed by 2000	Price proposed to be based on current costs
United Kingdom	From July 2001. Price likely to be about 13€/month	OFTEL will set price based on forward-looking long-run incremental cost

Source: European Commission (1999c).

a. Access to one copper pair.

b. Includes monthly rental of unbundled copper pair, where available, exclusive of value-added tax.

legal basis for the provision of unbundled access to the high-frequency spectrum of the local loop. As far as high-speed bit-stream access is concerned, there was no requirement in the legislation to have cost-orientated prices at the wholesale or retail level, although such prices would be subject to the normal provisions for competition law.

Accordingly, in 2000 the Commission published, and the European Parliament and Council adopted, a regulation on unbundled access to the local loop, which required member states to introduce it by the end of that year.[10] Enforcement action was threatened against the United Kingdom, which was not able to meet the target.

It is interesting (in the light of the discussion below about the relation between communications and competition law) that the Commission's analysis, while noting that the proposed recommendation would not in any way abridge the force of competition law, adopts a rationale for local loop unbundling that makes no appeal to standard competitive analysis. Under that competitive analysis, the natural way to proceed would be to define the market, identify dominance, and consider whether specific forms of behavior are an abuse of dominance. A natural outcome of this analysis might be the following:

—The relevant market is the market for the provision of access to the main distribution frame and the twisted copper pair leading to the subscriber's home.

—In particular geographic areas, the historic operator may have a degree of market power that goes beyond simple dominance and may extend through what is sometimes called superdominance to de facto monopoly. This situation may be due to inherited monopoly advantages, including vertical integration, monopoly of the provision of infrastructure, dominance in the provision of services, ubiquity, and brand awareness.

—Barriers to entry in the local loop are such that market power is likely to be nontransitory.

—In such circumstances, refusal to grant unbundled access to the local loop may constitute an abuse of the dominant position; it has the effect of eliminating a competitor's ability to compete on downstream markets with the owner of the facility; it involves tying access to the main distribution frame and the copper wire with access to switching capacity; it also has the consequence of limiting markets and technological development, preventing the emergence of new services.

A corollary of this analysis is that there may be circumstances in which unbundling of the local loop is inappropriate, because the competition in the local loop among rival infrastructure providers is already adequate.

Universal Service. The notion that a country's telecommunications operator should have a universal service obligation (USO) has a long history, becoming well established in the period of monopoly provision by either a public or a private operator. In telecommunications, universal service is widely recognized as having at least three dimensions:

—Universal geographical coverage, requiring that a network is rolled out over the whole of the territory in question, so that service is available for the whole population;

—Geographic averaging of tariffs, requiring that customers in the same category (for example, domestic or business customers), are charged the same tariffs, irrespective of their individual costs of supply; and

—Basic telecommunications service made available at an "affordable" price, so that no household (more realistically, few households) are denied access to the service. This is increasingly replaced by the more nuanced proposition that certain households, characterized by attributes such as low income or by their pattern of use of the telecommunications network, should have special "social" tariffs available to them.

The rationale for having a universal service obligation is based on a mixture of political, social, and economic considerations. It is desirable on political grounds that citizens in a democracy have access to the communication facilities that they require to exercise their political rights, and it is desirable on social grounds that all individuals have access to communication facilities, to avoid the emergence of a gulf between "information-rich" and "information-poor" groups. Economic arguments derive from "network externalities" and the link between network expansion and economic growth.

The notion that a universal service obligation for telecommunications should be sustained has survived the introduction of competition in the sector. It had previously been thought that competition and the USO were incompatible, because competitors would focus on high-value customers, defined by their calling patterns or their (low cost) location, leaving low-value customers to the operator with the universal service obligation. As competitors aggressively encroached on profitable markets, the commercial prospects of the USO operator would gradually unravel, jeopardizing the universal service obligation. However, further examination quickly revealed that it was possible to avoid this outcome if the costs of the universal service obligation were shared on some equitable basis between the operator discharging it and competitors. This triggered a discussion of appropriate means of estimating the cost and appropriate means of sharing it. A number of the governments and regulatory agencies have implemented procedures estimating the costs and in some cases have instituted arrangements for sharing the costs.

These developments have encouraged a new view of the universal service obligation and its funding as a *sui generis* tax and subsidy regime. Effectively, a limited amount of revenue is raised within the sector through "taxes," in the form of a percentage of operators' revenue or of a tax on particular services, such as interconnection, traded at the wholesale level. The revenue thus raised is then distributed to ensure that the universal service obligation is satisfied. This could in principle be done either by subsidizing particular groups of customers, to whom, for example, vouchers could be distributed, or by subsidizing operators providing services to particular customers. The former model might seem appropriate when the obligation relates to particular categories of customers, but it is less suitable when a net cost of the USO arises because of geographical cost differences.

The seventh and eighth recitals of the interconnection directive note that the concept of universal service must evolve to keep pace with advances in technology, market development, and changes in user demand. They then go on to note that member states may have more than one firm with universal service obligations and that it may be appropriate in due course to consider whether the obligations should extend to the provision of higher capacity services. The costs of universal service obligations should be calculated on the basis of transparent procedures, and when the USO represents an unfair burden on a firm, it is appropriate for it to be shared, provided that the sharing mechanism satisfies the principles of EU law, especially those of nondiscrimination and proportionality.

The approved procedure for costing universal service is adumbrated in the text and in an annex to the directive. Where networks are fully developed, the calculation should be of net costs attributable to services provided to customers who would not be served under normal commercial conditions. In areas with developed networks, the cost calculations should be based on the costs associated with serving those users, which would presumably be net avoidable costs. Where the network is still being developed, the cost calculation should be based on net incremental costs.

As a result of the directive, a number of studies utilizing the net avoided costs approach have already been carried out by NRAs or by the Commission. A key difficulty in undertaking such studies is that, for reasons of practicality, the areas within any country have to be allocated to a limited number of categories characterized by particular cost conditions. Yet it is one of the principles of USO costing that aggregation or averaging must inevitably reduce the net cost estimate, because it allows canceling out.

Second, it is difficult to attribute to customer groups profits associated with incoming calls to those subscribers, because incoming call data are often unavailable. Third, the computation of net avoidable cost hinges crucially on what assumption is made about the extent to which a subscriber notionally excluded from the network on cost grounds would continue to make calls. The implicit assumption of most studies is there would be no substitutional replacement. This is the assumption that minimizes the estimate of net avoidable costs. Estimates of the net cost of universal service obligations in member states tend to cluster around 2 or 3 percent of fixed telephony revenue—lower in the United Kingdom and higher in less densely populated countries, including France and some Nordic countries. In any event, only two member states have chosen to set up a sharing fund (France and Belgium), and they have been the subject of an investigation by the Commission's Competition Directorate.

After the launch of the eEurope initiative, however, universal service seems to be taking on a rather different meaning for the European Commission, which has stated the objective of connecting all schools and classrooms in Europe to the Internet by 2001. This is a deserving objective, but its attainment will impose substantial costs. It is currently unclear how they will be met.

Thus we could say that the Commission has made a step away from a vision of universal service based on a "merit good" argument, close to the French notion of *service public*: it was thought relevant, but in practice achievable through market mechanisms. The new approach is much closer to a view of access as a generator of positive economic and social externalities.

Leased Lines. The leased-lines problem in Europe is clearly summarized by figure 3-1.[11] Leased lines are far more expensive in Europe than in the United States: the standard example is that a line between London and Paris is more expensive than a line between London and New York. True, prices slowly fell in Europe throughout the 1990s, but they fell in the United States too.

Why are leased-line prices so high, and why are they only slowly converging to international levels? This is an interesting problem. Most of the regulatory measures that were necessary to make prices drop have been part of the EU liberalization package since 1996, but nothing was done until 1999. The standard explanation for the delay provided by the Commission is that NRAs had "difficulties in acquiring sufficiently detailed cost data from incumbent operators."[12] This response is unconvincing. Leased lines have traditionally been run as a separate business by operators, and cost accounting problems should have been much smaller than in other areas—such as interconnection—where regulatory intervention has been extensive.

An alternative explanation is that leased lines were a substantial cash cow for the incumbents, which were probably obtaining very high rates of return from them. A high price for leased

Figure 3-1. North American and European Intercity Leased-Line Price Offers, Early 1999

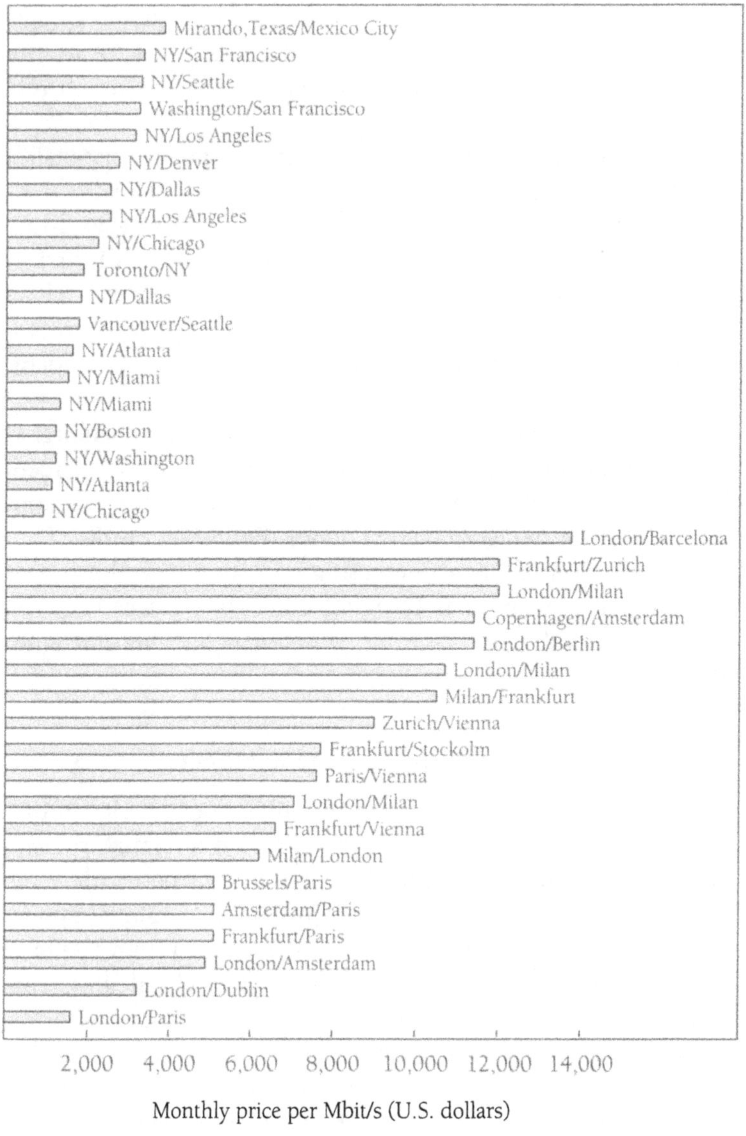

Mirando,Texas/Mexico City
NY/San Francisco
NY/Seattle
Washington/San Francisco
NY/Los Angeles
NY/Denver
NY/Dallas
NY/Los Angeles
NY/Chicago
Toronto/NY
NY/Dallas
Vancouver/Seattle
NY/Atlanta
NY/Miami
NY/Miami
NY/Boston
NY/Washington
NY/Atlanta
NY/Chicago
London/Barcelona
Frankfurt/Zurich
London/Milan
Copenhagen/Amsterdam
London/Berlin
London/Milan
Milan/Frankfurt
Zurich/Vienna
Frankfurt/Stockolm
Paris/Vienna
London/Milan
Frankfurt/Vienna
Milan/London
Brussels/Paris
Amsterdam/Paris
Frankfurt/Paris
London/Amsterdam
London/Dublin
London/Paris

2,000   4,000   6,000   8,000   10,000   12,000   14,000

Monthly price per Mbit/s (U.S. dollars)

Source: OECD (1999).

lines was also welcome for entrants, which could charge high prices even to large customers. It would thus seem that there was a community of interest among operators to have high leased-line prices: final customers, at least before the massive expansion of Internet penetration, were not much aware of the situation, and in any case their market power was much less than that of the operators.

Worries about Internet penetration in Europe, the expansion of capacity stimulated by high prices, and a more vocal regulatory stance, have engineered a fast fall in European leased-line prices since 1997. Thus the Competition Directorate of the European Commission has opened an inquiry, which may well lead to an antitrust procedure. The Commission also issued, in November 1999, a recommendation to NRAs to investigate operators if charges exceed benchmark values.[13]

All in all, price decreases in percentage terms have been higher on European routes than on transatlantic ones (by 75 percent on the London–New York routes, 88 percent on the London–Paris routes, 84 percent on the London-Frankfurt routes).[14]

So the combination of increasing competition and stronger regulatory intervention is bringing leased-line prices down in Europe, although it will be some time before they become comparable with U.S. prices.

Implementation of the Package in Mobile Services

Mobile telephony is the success story of European telecommunications and a rare case when most policy and regulatory decisions turned out to be right. The first correct policy decision was the development and adoption of the GSM (global system for mobile) standard. As is well known, European countries have a poor record of reaching agreements on new technologies. Generally, each one pushed the solution proposed by its own industry. In the 1960s, for instance, it was impossible to reach a common decision about color television standards; attempts to agree to a high-definition television (HDTV) standard also ended in failure.

A GSM standard was, however, adopted. The reasons are discussed at length by Jacques Pelkmans: a strong commitment by a sufficiently large group of national monopoly operators, mostly in state hands, that stood to reap the gains from the diffusion of a European standard that would offer smooth continental roaming; and effective political support, given the strong European connotation of the project.[15] The GSM project also gained momentum from the substantial commercial success that was met by existing analog services, available in most countries at the beginning of the 1990s.

Licensing. GSM licenses were allocated in the middle of the 1990s. All fixed-line incumbents got one, and the European Commission exercised considerable pressure to have at least two licenses in each country. This led to a rapid increase of the number of firms. At the end of 1999 there were thirty-seven operators with a GSM license in the 900 MHz band, two per country except for France, Italy, Portugal, Finland, and Sweden, which had three. Thirty-four operators had nationwide DCS 1800 licenses (the European flavor of PCS 1900), and there were still fifteen operators with analog licenses.

Almost all the licenses are nationwide—the natural choice in a continent of national incumbents and (relatively) small land areas. This helped to foster penetration and contrasts with the U.S. decision to allocate spectrum to several hundred separate areas.

It is envisaged that the Commission, in conjunction with NRAs, will regularly determine the relevant product and service markets on the basis of which SMP will be assessed. GSM licenses were granted on the basis of competitive applications, or "beauty contests," and recipients mostly paid nominal fees. Thus, the rent stayed with them, instead of accruing to the taxpayers. The optimality of such an outcome could be discussed at length: it is fair to say, however, that mobile operators in Europe have by and large made use of their position to roll out very quickly good-quality networks and achieve in all countries a very high degree of coverage (never below 95 percent in terms of population).

Penetration rates have passed the 50 percent mark in the Nordic countries and in Italy and are substantially above the U.S. mark in most large European countries.

The age of cheap licenses came abruptly to an end in Europe after the United Kingdom's decision to auction 3G spectrum for what are known as UMTS services in Europe (or IMT-2000). As is well known, prices far exceeded expectations and brought to the Exchequer a total of 35 billion euro. Subsequently, total expenditure for licenses in the German auction equaled 50.4 billion euro. Although prices were lower in other countries, the total cost to operators of the current UMTS significantly exceeded 100 billion euro.

The rationality of the bids was disputed in some quarters, as UMTS is a new technology whose appeal to mobile users will depend on several factors (such as content availability and the price reaction of GSM competitors) that are very difficult to evaluate.

There is something rather counterintuitive in the notion of outsiders, including Commission officials who have criticized the size of the bids, having a better appreciation than operators themselves of the value of 3G licenses, which clearly includes the value of protecting existing 2G revenues. The degree of technological uncertainty is, however, very high, while the cost of borrowing to pay for the licenses is increasing the companies' cost of capital overall.

In any case, since nothing so dramatic seems likely to happen to the price of the available spectrum in the United States, it will be interesting to see if any differential effects arise in the two areas from the large differences in the cost of doing wireless business in Europe and America.

Interconnection and Final Price Regulation. The issue of interconnection in mobiles has received little attention in Europe. A relevant (but largely unspoken) policy decision was indeed the one to grant ample pricing freedom to wireless providers. Typically, mobile-to-fixed (M2F) prices had been set by the regulators for analog systems. More freedom was granted to retail GSM mobile-to-fixed prices, given the existence of (at least) one GSM competitor, plus the continuing existence of analog systems. Even more freedom was granted to mobile-to-mobile (M2M) call pricing—typically agreed on between operators. The same relative freedom applied to fixed-to-mobile (F2M) prices. This was crucial, as in Europe such calls were very expensive but were paid by the fixed-line callers.

We believe that this regulatory framework for mobiles (including its obvious imperfections) was instrumental in delivering the massive growth of mobile penetration witnessed in Europe. Very often, calling parties (who are the ones who pay, under the European model) did not know what prices they were being charged. These were usually high but of course had no direct impact on mobile subscribers. Indeed, it is arguable that in some countries F2M calls were subsidizing fixed charges and M2F calls. Many European mobile operators were competing on M2M and M2F prices, while keeping F2M prices at a high level. This could be rationalized as a case of Ramsey pricing.

F2M calls inevitably became more of an issue as market penetration grew, and numerous mobile users found out that their fixed telephone bills contained a lot of such traffic generated by sons, daughters, and relatives. Firms found this out rather earlier. After a UK Monopolies and Mergers Commission Inquiry in 1998, and a similar investigation carried out by the Competition Directorate of the European Union in 1998 and 1999, F2M interconnection and retail prices began to drop substantially in Europe.

Looking Ahead: The 1999 Communications Review

In November 1999, the Commission published its Communications Review, intended to lead to a legislative framework to come into effect starting in 2003.[16] The review occurs fifteen years after the first step toward liberalization in the field of equipment and ten years after the first step in services, and thus has the benefit of perspective. It was released by a newly formed Commission and thus provided a good opportunity to set out a strong policy statement.

The review proposed a legislative framework that will come into full effect in 2003. It has therefore a medium- to long-term perspective—a hard task in an industry as dynamic as telecommunications. One of its major challenges is to combine a vision of the future with some policy specifics.

The review set out three major policy objectives: to promote an open and competitive European market for communications services and equipment; to benefit the European citizen; and to consolidate the internal market in a converging environment. It is hard to disagree with these general aims. The problem is what policy consequences to draw from them.

The review is a complex document, in which broad policy statements are intertwined with technical points, which, of course, are often very important. Here, we summarize the major points from a broad perspective, stressing strategy rather than detail.

The review starts out by stating that the existing legislative framework was originally designed to create a competitive market. A new framework is now needed to manage new, dynamic markets, where competition is growing. Hence, ex ante regulation of dominant market players should continue but be progressively reduced as competition develops. New markets should be subject to a light regulatory touch.

The new regulatory framework should be built on five principles. Regulation should be kept to a minimum to achieve those objectives, be based on clearly defined policy objectives, and enhance legal certainty. Regulation should also aim toward technical neutrality and—once put in place at the global, European, or national level—be enforced at the lowest possible level in obedience to the principle of subsidiarity.

The new proposals are to be enshrined in a framework directive and will be enacted though a broad set of instruments, including directives on licensing (or authorization), access and interconnection, universal service, and privacy and data protection. Five draft directives were published in July 2000 for consideration by the European Parliament and Council.[17]

The framework directive looks much like a European Tele-communications Act. It specifies policy objectives to be pursued by each member country using the methods of its choice; consolidates the powers and responsibility of the national regulatory authorities, including important new procedures for a right of appeal covering not only procedures adopted by the NRA but also the facts of the case.

One of the main provisions of the draft directive concerns undertakings with "significant market power" (SMP). That phrase had previously been applied in the interconnection directive to operators that had a share of some prespecified and broadly defined market in excess of 25 percent. These operators had requirements imposed on them, such as separate accounting and the obligation to supply at cost-orientated prices. The 1999 Communications Review envisaged the maintenance of the 25 percent market share threshold in the new regime; the draft directive retains the SMP terminology but redefines it in terms equivalent to the standard European Court of Justice definition of dominance (the ability to behave to an appreciable extent independently of competitors, customers, and ultimately consumers), normally triggered by a market share of the order of 50 percent. Special mention is also made of problems associated with the leveraging of market power by vertically integrated firms.

However, unusually, it is envisaged that the Commission will regularly determine, with NRAs, the relevant product and service markets on the basis of which SMP will be assessed. It will publish guidelines on market analysis and the calculation of significant market power to be implemented by the NRAs.

This vision represents a welcome departure from the previous proposal, but the effectiveness of the regime will depend considerably on how markets are defined and how market power is assessed. If narrow market definitions are employed, many firms may be found to have SMP, especially if the analysis of emerging markets fails to consider the essentially transient advantages enjoyed by first movers in the market for many telecommunications services. It is also interesting to note that the procedure involved the ex ante application of analytical techniques (market definition and the identification of dominance) that are conventionally adopted on an ex post basis. The operation of these procedures is described in more detail in other directives, such as that for access.

Other features to emerge from the 1999 review and subsequent draft directives are as follows:

Licensing. The Commission proposes general—rather than specific—authorizations in both telecommunications and broadcasting, reserving the latter instrument for cases in which scarce resources (for example, spectrum) are to be allocated. Authorizations should be tradable. A transition to Europewide licenses is envisaged. The fact that in EU countries the practical allocation of rights of way is chiefly in the hands of local authorities, which may have an interest in restraining entry, is mentioned, although no specific remedy is discussed.

Rebalancing of tariffs. The Commission is aware that this is at best incomplete in several countries, but it does not address the prospect that unbalanced tariffs may encourage inefficient entry in the long-distance market. A stronger emphasis on rebalancing, with abatements for particular groups in the interest of universal service, is a better solution.

Leased lines. These seem largely to escape the Commission's attention in the review, but this may be a wrong impression. The ineffectiveness of the directives relating to leased lines seems to be recognized implicitly, but much hope is pinned on a subsequent recommendation, which we have discussed above.

Spectrum. The Commission plans to do little about how the member countries allocate spectrum: beauty contests and auctions are likely to coexist for a while. Significantly, secondary trading of spectrum will be explored, with a view to making it legal (an overdue reform) but not mandatory.

Universal service. Definition of its scope at the EU level is discussed as follows. The financing of universal obligations has so far not been too problematic. Only two countries have created a universal service fund, and both are subject to an infringement procedure. It may, however, become a problem in the future, as the market share of the incumbent falls below a certain threshold and such financing will be kept under review. The Commission recommends consideration of "pay or play" schemes.

Affordability. The Commission supports the general idea and rightly makes it a responsibility of member states. But it does refer to the need to set out "clear pricing principles at a European level" in order to ensure affordability.[18]

Price transparency. The review mentions this issue almost in passing and proposes a significant innovation: that of providing all customers with call-by-call tariff information.

Internet. All in all, the Internet is to remain free from regulatory intervention. Providers will not need authorizations. If and when voice-over Internet services qualify as voice telephony services, they will require a general authorization.

Looking Ahead: Crucial Issues

Rather than offer a detailed criticism of the Commission proposals, we think it more useful to turn to some of the background issues that are likely to affect regulation of the sector, such as enforcement, related institutional problems, and the balance between regulation and the use of national or European competition law.

Enforcement. A regulatory system is only as effective as it is enforceable, and a blind faith in the willingness of member states to fulfill their duties under the treaty is at best optimistic, at worst misguided. Indeed, the implementation record of EU legislation in several member states has been poor.[19]

The services directive was implemented years late in Spain, Italy, Denmark, and the Netherlands, and the Court of Justice has recently issued rulings against a number of countries including Greece for their failure to implement the 1992 directive on leased lines. As noted above, leased lines are a flagrant demonstration of the problems of enforcement. There are specific directives covering them that have liberalized their provision since July 1996 and imposed an obligation upon SMP operators (that is, all European incumbents) to supply at cost-oriented prices. Nothing, however, happened until recently.

How can European-level legislation be enforced within member countries in a fast and efficient way? This is the Holy Grail of telecommunications liberalization in Europe. On paper, this should not be too difficult for the following reasons:

—Directives approved by the European Parliament are effective within member states even if they have not explicitly been transposed into international legal systems through national laws. National courts must take them into account.

—The Commission has powers to promulgate a directive whenever it deems necessary to prevent or remedy infringements to the competition provisions of the Treaty, whether in the field of an abuse of dominant position or of illegal state aid to companies. Indeed, some of the past directives that have been the most effective have been of this kind.

—The Commission may start infringement procedures against countries that do not abide by directives.

The wheels of civil justice grind often too slowly for communications issues, and so legal remedies do not work in several cases. The compliance mechanisms at the European Union level are weak: a fine can be imposed by the Court of Justice only after the Commission has brought infringement proceedings against the offending country twice, and this takes years. Aggrieved "entrants" may be able to rely on nonimplemented directives against an operator when the operator is classified as an "emanation of the state," although damages are not necessarily available when these are not recognized under national law. Finally, the possibility for operators to obtain damages from a member state through its failure to respect its Treaty obligations does exist, but the conditions under which this may be possible are unclear and require clarification by the European Court of Justice. Unless a way forward can be found, the objectives set out in the new proposals may only be achieved patchily.

In such a landscape, the Commission has often relied on a rather indirect approach. As we have seen in the case of interconnection and leased lines, this has been based on recommendations, which are no more than their name implies and are therefore not legally binding. Typically, a recommendation would state that the Commission has grounds to believe that a price (say, of double transit interconnection) is too high, and "recommends" that NRAs have a close look at the accounts of the incumbent if the price of such service falls above a "ceiling" calculated on the basis of the three lowest prices for double transit interconnection to be found in any EU country.

Such recommendations are carefully crafted. The proposal is presented as a temporary device, to be employed only until better data are available. In several countries this task may take some time, so that prices will be regulated for a while with reference not to cost but to a somewhat arbitrary price ceiling set by the Commission. It is an open question whether such regulatory decisions could be successfully challenged in court.

The Institutional Problem. The enforcement problem would look less serious if some form of regulatory centralization could be achieved. Could this be changed by the creation of a European-level Regulatory Authority (ERA)? This is a recurring theme in the European debate. It officially surfaced for the first time in the early 1990s, when concern was growing among operators and Brussels officials about the slowness with which member countries were implementing the first open network provision (ONP) directives. The Commission's Directorate concerned with telecommunications commissioned a study of the feasibility of an ERA. The study found broad support for the idea coming from industry but concluded that any such body would require an amendment to the European Treaty (a very cumbersome process involving not only approval by national governments, but referendums in several countries).[20]

The issue is now surfacing again in the academic literature, where it is argued that heterogeneous local rules favor local incumbents and slow the pace of liberalization; deregulation is easier if regulation is centralized; moreover, Internet growth is making the very concept of "local" obsolete.[21]

While it is easy to agree with this analysis, it should be pointed out that institutional considerations are of paramount importance:

unless an ERA took substantial powers away from the NRAs (and this cannot happen unless the European Treaty is modified), the regulatory framework in Europe could become more complex, with a growing risk of conflict among different bodies.[22]

The legal notion of subsidiarity, which sits at the core of the European Treaty, is often invoked. This concept can be stated simply: let an issue be dealt with at the lowest possible level within the Union. Obviously, the concept may lend itself to instrumental use in order to protect national vested interests.

The text of the Treaty is, however, clear, and limits any instrumental interpretation: article 3B states that "the Community shall take action, in accordance with the principle of subsidiarity, only if and in so far as the objectives of the proposed actions cannot be sufficiently achieved by the member states and can therefore, by reason of the scale or effects of the proposed action, be better achieved by the Community."

The article can be given a straightforward economic interpretation by stating two necessary conditions for Community action in areas that do not fall into its exclusive competence:

(a) the existence of scale effects in some areas, creating a situation in which the individual action by a member state would be inadequate because it would entail excessive costs or provide reduced benefits;

(b) the existence of relevant effects that go beyond the frontier of that particular country and affect other member countries or the Union as a whole.[23]

In the case of telecommunications, (a) probably applies in many cases. Instances in which (b) could be relevant abound in the information industries. Nonetheless, subsidiarity has often been invoked whenever the Commission has tried to take away regulatory powers from member states. This is the basic reason why NRAs have far greater powers than any European-level body in telecommunications regulation.

Unsurprisingly, the Commission has avoided proposals for a major centralization that are unlikely to command support among the member states. It proposes instead the creation of a High-Level Communications Group (HLCG) composed of the

Commission and NRAs, stemming from the current High-Level Regulators Group. This group would work with European-level bodies such as standards organizations and industry representative groups. It would concentrate on "assisting the Commission in maximising uniform application of national measures adopted under the regulatory framework laid down in community legislation." According to the draft directive, this assistance would include

—Examining national measures adopted under the framework directive to promote uniformity of application;

—Adopting agreed positions on the detailed application of legislation, thus facilitating pan-European services;

—Assisting in the decisions on market definitions noted above;

—Endorsing codes of practice associated with community legislation; and

—Considering issues brought up by member states.

In addition, there would be a communications committee, which would replace the ONP and licensing committees and advise the Commission on draft measures.

This sounds much like business as usual, except for the extent to which the Commission's freedom of maneuver is limited. The role of these groups in enforcement is likely to be limited; hence, it is probably not enough.

Ex Ante and Ex Post Rules. Quite apart from such general issues, the remedies envisaged by the Commission to deal with the already existing access issues seem to rely too much on a regulatory framework that has clearly shown several shortcomings and too little on a full deployment of its powers under competition law. This is all the more surprising, as a major step in this direction has already been taken by the Commission, in the access notice published by DG Competition (then known as DG IV) at the beginning of 1998.[24]

This notice set forth an important principle: the market for telecommunications is, in general, the EU as a whole. Hence, Commission antitrust powers should fully apply to what had previously been considered purely national markets.

Until 1998, DG IV's activities in telecommunications had concentrated on the vetting of agreements, joint ventures, and mergers under articles 81 and 82 of the EC Treaty. These have all been approved. Only one case—in 1997, against Deutsche Telekom—was concerned with the abuse of dominant position.

Under (new) article 86, very few cases were reported to the Commission. They were mainly based on discriminatory actions by the administration of a member country against new entrants. In two cases, a formal action was taken against the government; in others corrective measures were proposed by the member states to remedy the distortions.[25]

With the beginning of "full competition" in Europe and with the liberalization of alternative infrastructures since January 1, 1998, the Competition Directorate has stepped up its action on telecommunications issues. It has started monitoring implementation of liberalization and harmonization directives, as well as issuing policy notices on topics related to telecommunications (access, voice-over Internet, and the separation of cable and telecommunications activities). On the basis of such monitoring, the Commission has opened several infringement proceedings against member states that failed to comply adequately with the directives. It conducted a survey in 1998 on the fixed-to-mobile termination charges, pushing operators and NRAs to act on the matter, and carried out a similar inquiry into the cost of leased lines.

The most far-reaching action undertaken by the directorate was, however, the release in March 1998, of its "Notice on the Application of the Competition Rules to Access Agreements in the Telecommunications Sector."

The definition of "access" chosen is very broad, and it includes not only interconnection but also access to any facility necessary to provide service to end users. Thus, the Commission's powers of intervention in *purely national* access and abuse of dominance issues are clearly established for the first time in this notice: articles 81 and 82 can be directly applied even if the companies involved operate within the same member state.

The Commission is politically correct about subsidiarity. It states in the notice that ONP-related procedures (that is, ones basically undertaken by the NRAs) should resolve access problems in the first place at a decentralized, national level. It suggests the further possibility of a proceeding at Community level in certain circumstances. It also makes clear, however, that the NRAs must ensure that actions taken by them are consistent with EC competition law.

If the application of the principles set forth in the notice is consistent, and they stand up in court, and also if the directorate has enough staff to handle a growing telecommunications-related workload, the construction of a "coherent legal and regulatory framework for the Information Society in Europe" (a primary goal set forth by the Bangemann Report in 1994 but always elusive) would take a decisive step forward. An interesting development in Germany will provide an early test of the Commission's resolve: Arcor appealed in July 1999 to the Commission's Competition Directorate about the level of rates set by the German Regulatory Agency for access to the local loop. In a different case, which however points in the same direction, Worldcom has complained about fixed-to-mobile termination charges levied by ten operators; the Commission is taking forward some of these complaints.

Conclusions

On balance, European telecommunications liberalization has been unexpectedly successful. When it was initiated, all the incumbents in member states were state-owned, inefficient monopolies; they all had powerful lobbies through which politicians, trade unions, and suppliers combined their strength to avoid change. Prices were very high, and their structure implied massive cross-subsidies. In a relatively contained span of time, considerable results have been achieved: several operators have been fully privatized (with the notable exceptions of the French and the German incumbents), entry has been liberalized, and prices have dropped. The mobile sector has fostered very high growth rates, which have given Europe a worldwide lead.

All this has been taking place within an EU governed by layers of supranational and national legislation, with a nonelected governing body (the European Commission) as a driving force. This complex—and often litigious—institutional framework has given rise to problems in the implementation of EU directives at the level of member states. (The United States has experienced similar problems within a different federal framework.)

Overall the success is striking: according to the Commission, fixed-line tariffs are now lower than in the United States for distances above 50 km and 200 km (see figure 3-2).[26] Interconnection rates in some countries are lower or at the very least comparable with the United States, and Europe also enjoys a positive competitive advantage in mobiles. All this has been obtained in a very few years, so it is reasonable to ask: is it sustainable? We would suggest caution, since the structural characteristics of the European telecommunications industry have in some way been biased by the liberalization framework and its problems of implementation toward a service-based competition model.

Let us consider first the infrastructure issue. It is difficult to provide comprehensive proof of a bias against the construction of competing infrastructure, but inferences can be drawn from data on network deployment and prices. Such data are in scarce supply, but a recent report to the European Commission has usefully put together data relating to Pan-European network deployment by major new entrants on continental routes.[27] These are patchy at best.

We have no consistent data about local deployment of infrastructure by new entrants in Europe, but nonsystematic evidence suggests that it is quite limited, being concentrated—in continental Europe—in the larger business centers, such as Brussels and Frankfurt. High leased-line prices also offer indirect evidence of the insufficient deployment of new infrastructure by entrants.[28] In the local loop, the market share of all operators apart from the largest in the provision of lines is less than 10 percent

Figure 3-2. Actual Price of Three-, Five-, and Ten-Minute Regional and Long-Distance Calls in the European Union and the United States, August 1999

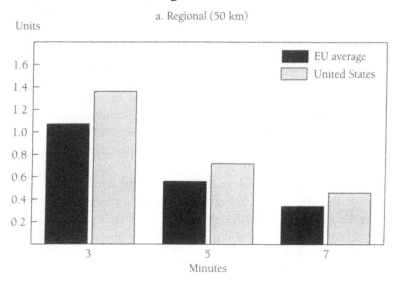

a. Regional (50 km)

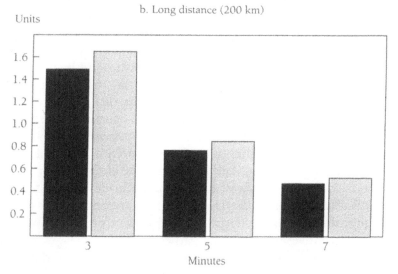

b. Long distance (200 km)

Source: European Commission (1999b).

throughout the European Union, except for the United Kingdom and Finland.

In a way, the service bias in Europe is unsurprising, if we consider that existing legislation was drafted in the first half of the 1990s, when—outside the United Kingdom—most incumbents were state owned. The idea of having competition in services on a big, publicly owned network was then quite attractive.

If insufficient competing infrastructure development is a major problem, then we do not think that the remedies proposed by the Commission in its draft directives go far enough: indeed, they may increase such bias. Further, we believe that such remedies do not make full use of the powers available to the Commission under competition law.

Under the guise of neutrality between infrastructure and service competition, the Commission seems to be more preoccupied with opening existing infrastructures than encouraging the construction of new ones. Thus entry is vigorous and prices have been dropping fast, but this does not seem to provide medium-term incentives for the extensive deployment of alternative infrastructures. In particular, it is unproven that a pervasive local loop unbundling approach, if based on cost-based pricing, would provide the necessary incentives in Europe for massive broadband investment by both incumbents and newcomers.

The drop in prices may also be a pointer toward ineffective, rather than effective, regulation, insofar as rate rebalancing has been at best an uneven process across Europe. By and large, the retail rate structure is still far from the cost structure. Line rentals are too low, while long distance is too expensive. New entrants get access to incumbents' networks at cost-related interconnection charges and then exploit arbitrage opportunities between those charges and the unbalanced long-distance tariffs. This is all right as an asymmetrical, pro-entry provision. But it cannot go on for long, since it provides incorrect investment incentives to all relevant parties.

These issues are not adequately tackled in the draft directive, but no coherent European policy for the Information Society can be developed without a clear long-term option that favors the construction of new, broadband networks. Mandatory local loop unbundling might be implemented in some countries in a way that hindered investment and technological upgrading.

In summary, a combination of regulatory instruments is needed that will provide a flexible framework within which to resolve the many issues that the development of broadband and multimedia services will give rise to in the near future. These should involve the national application of a limited set of ex ante rules determinted at the EU level, combined with ex post vetting based on the application of European competition law within national telecommunications markets.

Notes

Chapter 1
Telecommunications Policy in
North America and Europe

1. A few countries were given derogations.

Chapter 2
Telecommunications Policy Reform in
the United States and Canada

1. The Telecommunications Act was signed into law by President Bill Clinton on February 8, 1996.

2. See *MCI Telecommunications Corp.* v *FCC*, 561 F.2d 365 (D.C. Cir. 1977), *cert. denied*, 434 U.S. 1040; and *MCI Telecommunications Corp.* v *FCC*, 580 F.2d 590 (D.C. Cir. 1978), *cert. denied*, 439 U.S. 980 (1978). Liberalization of terminal equipment, although contested by state regulators, proceeded far more rapidly in the late 1970s.

3. The FCC preempted state rate regulation in 1986, but a federal appeals court narrowed this preemption to wireless *interstate* services in 1987. *NARUC* v. *FCC*, No. 86-1205 (D.C. Cir. 1987).

4. The 1984 consent decree created 161 local access and transport areas within which the divested RBOCs could offer long-distance service. However, the RBOCs could not provide service between LATAs, even those in the same state or wholly within the RBOC's region. In effect, this limited the RBOCs to service areas comprising about 20 percent of all long-distance service.

5. See Hausman (1997, pp. 1–38).

6. P. L. 104-104, 110 Stat. 56.

7. *Telecommunications Act of 1996 Conference Report*, 104 Cong. 2d sess., Report 104-458 (Government Printing Office, January 31, 1996), p. 1.

8. See FCC web page, http://www.fcc.gov/Reports/telecom-rpt.html.

9. Telecommunications Act of 1996, sec. 253, "Removal of Barriers to Entry."

10. Telecommunications Act of 1996, sec. 251, "Interconnection," (3), (4).

11. See Hazlett (1999).

12. For further analysis of the act, including many of the regulatory issues we exclude from our paper, see Kellogg, Thorne, and Huber (1996).

13. Equal access refers to the ability of competitive carriers to offer interconnected services without discrimination. In long distance, it has generally meant that customers of nondominant firms not find it necessary to dial additional numbers or pay additional tolls in order to substitute carriers.

14. See MacAvoy (1996).

15. This is not an antitrust market because long-distance carriers offer both intra-LATA and inter-LATA services and thus compete to some extent with local-exchange carriers. Moreover, one might argue that large businesses and residential customers are in different markets. The Bell companies cannot compete in the inter-LATA market in most states yet, and long-distance carrier participation in intra-LATA markets was impeded by state regulators until 1999. Therefore, including local-exchange carriers' long-distance revenues—equal to 15 to 20 percent of all long-distance revenues—in the tabulations in table 2-2 would be misleading.

16. The HHI is the sum of the squared market shares for all market participants. An HHI of 2,500 is roughly equivalent to that generated by an industry with four firms of equal size.

17. Taylor and Taylor (1993).

18. See Robert Hall, *Affidavit on Behalf of MCI*, U.S. Federal Communications Commission, In the Matter of Application of SBC Communications, Inc., Pursuant to Section 271 of the Telecommunications Act of 1996 to Provide In-Region, InterLATA Services in Oklahoma (April 1993).

19. See MacAvoy (1996).

20. These are average prices for domestic interstate calls (United States) and for national calls (Canada).

21. Sprint Canada's rate. See www.fonorola.com.

22. See Crandall and Waverman (1996).

23. There are fixed marketing, billing, and administrative costs that must be covered from revenues, but these costs are increasingly being recovered on a flat monthly basis. Marginal prices should not reflect these costs, assuming healthy competitive market constraints.

24. Such entry has now been authorized by the FCC in five states: New York, Massachusetts, Texas, Oklahoma, and Kansas.

25. TNS, *Bill Harvesting* data.

26. This estimate was derived by Jeffrey Rohlfs and Robert Crandall in FCC comments submitted in support of the CALLS proposal in 1998. CC Docket nos. 96-262, 94-1, 99-249, and 96-45.

27. See Hausman (1997).

28. For a description and critique of this methodology, see Kahn (1998); Hausman (1999a, 1999b).

29. Jeremy Pelofsky, "Appeals Court Throws Out FCC Access Price Model," *Reuters*, July 19, 2000.

30. Many states have uniform wholesale rates for unbundled elements despite large cost differences that exist between dense areas and more rural areas. In table 2-3, the rates are for the most densely populated areas in the states that allow different rates across geographical areas.

31. See Crandall and Waverman (2000).

32. See Crandall and Hausman (2000). See also Hausman (1999a).

33. This abstracts from the possibility of implicit or explicit subsidies. That issue, while interesting, forms a separate discussion.

34. See the FCC's Common Carrier Bureau data on carrier revenues.

35. "Critics, Backers of '96 Act Joust on Phone Markets," *National Journal's Congress Daily*, March 10, 1999, p. 4.

36. Indeed, one of the four has been publicly listed only since March 1994. Many more firms were listed for some part of the sample period, including several firms that were delisted when acquired via merger. The CLEC firms listed are the only companies lasting for the entire five-year period (or close to it), 1994–98, on the website devoted to tracking competitive local-exchange carrier stocks: www.clec.com.

37. In early 1999, beta for Winstar was 1.57, Intermedia 1.26, GST 1.91, and ICG 2.56. The risk premiums associated with these betas account for up to about 80 percent of the excess CLEC returns.

38. Cellular Telecommunications Industry Association, *Semiannual Survey*, December 1999.

39. Telecom Decision CRTC 97-8, *Local Competition—Local Unbundling and Interconnection*, May 1, 1997.

40. Federal Communications Commission, First Report and Order and Third Notice of Proposed Rule Making Re: *Redevelopment of Spectrum to Encourage Innovation in the Use of New Technologies,* 7 F.C.C.R. 6886, September 1992.

41. See Merrill Lynch, *The Next Generation III*, March 10, 1999, p. 13, for data on the number of carriers by market.

42. See Hausman (1999c).

43. See the Declaration of Robert W. Crandall and Robert H. Gertner filed on behalf of Bell Atlantic in the merger of Bell Atlantic and GTE before the Federal Communications Commission, 1999.

44. Federal Communications Commission, *Annual Report and Analysis of Competitive Market Conditions with Respect to Commercial Mobile Services: Fourth Report*, FCC 99-136 (June 24, 1999), 12–15.

45. For a fascinating, detailed account of this episode, see Krasnow, Longley, and Terry (1982).

46. U.S. Department of Justice (1987).

47. See Huber (1997, p. 99).

48. It was not a universal political success, however. The chair of the Senate Commerce Committee and key sponsor of the act, Larry Pressler (R-SD), was the only incumbent senator defeated for reelection in 1996. His opponent raised the Telecommunications Act as a campaign issue.

49. Kahn (1998).

Chapter 3
The Liberalization of
European Telecommunications

1. See European Commission (2000a, 2000c).

2. For a comprehensive outline of telecommunications policy in the EU, see European Commission (1999c).

3. The key directives are the open network provision (ONP) framework directive 90/387/EEC, amended by directive 97/51/EEC; the leased-lines directive 92/44/EEC; the new voice telephony directive 98/10/EEC; the licensing directive 97/13/EEC; the interconnection directive 97/33/EEC; and the numbering directive 98/61/EEC. For a detailed but concise description of EU legislation, see Scherer (2000).

4. European Commission (1999d, p.12).

5. Pontiggia and Vandenbroucke (2000).

6. European Commission (1999d, p. 36).

7. See annex V of the interconnection directive.

8. See European Commission (1998a, 1999b, 2000b).

9. See European Commission (1999d).

10. European Commission (2000i).

11. Leased-line prices are notoriously hard to compare. The data from the Organization for Economic Cooperation and Development used in figure 3-1 are given by telecommunications capacity markets and thus are probably more significant for spot transactions. List prices, however, are often of little relevance because of widespread discounting. What we need is a broad idea of price differentials across the Atlantic, and the data are certainly good enough for this. See OECD (1999).

12. See European Commission (1999b, p. 7).

13. The benchmarks were, for a 34 Mbit/s circuit: 1.800 euro/month for 2 km, 2.600 euro/month for 5 km; for a 2 Mbit/s circuit: 350 euro/month up to 5 km; for a 64 kbit/s circuit: 80 euro/month up to 5 km. See European Commission (1999a).

14. Prices referred to are from Band-X indexes, relating to E1/T1 lines.

15. Pelkmans (2000).

16. See European Commission (1999d).

17. European Commission (2000d, 2000e, 2000f, 2000g, 2000h).

18. See European Commission (1999d, para. 4.4.3).

19. For an early assessment, see Cullen and Blondeel (1994).

20. The report was published much later as National Economic Research Associates (NERA) (1997).

21. See Kiessling and Blondeel (1998); Lehr and Kiessling (1998).

22. See Eurostrategies–Cullen International (1999).

23. See Pelkmans (1990).

24. See European Commission (1998b).

25. See the discussion of Prosperetti, Cimatoribus, and della Torre (1999).

26. A local call price comparison is not significant, given the large differences in the rate structures, mostly flat in the United States and usage-based in Europe.

27. See Logica Consulting (2000).

28. This has been recently recognized by the European Commission in its leased-lines recommendation. See European Commission (1999a).

References

Crandall, Robert W., and Jerry Hausman. 2000. "Competition in U.S. Telecommunication Services: Effects of the 1996 Legislation." In *Deregulation of Network Industries: What's Next?* edited by Sam Peltzman and Clifford Winston, 73–112. Brookings.

Crandall, Robert W., and Leonard Waverman. 1996. *Talk Is Cheap: The Promise of Regulatory Reform in North American Telecommunications.* Brookings.

———. 2000. *Who Pays for Universal Service? When Telephone Subsidies Become Transparent.* Brookings.

Cullen, Bernard, and Yves Blondeel. 1994. "Union Measures Taken in the Telecommunications Area and Results Achieved." In *Implementing European Telecommunications Law* conference. Brussels: European Commission (November).

European Commission. 1998a. "Interconnection Tariffs in Member States as of 1st December 1998." www.ispo.cec.be

———. 1998b. "Notice on the Application of the Competition Rules to Access Agreements in the Telecommunications Sector." 98/C 265/02. Published in the *European Union Official Journal,* August, 22, pp. 2–28.

———. 1999a. "Commission Recommendation on Leased Lines Interconnection Pricing in a Liberalised Telecommunications Market." www.ispo.cec.be

———. 1999b. "Fifth Report on the Implementation of the Telecommunications Regulatory Package." www.ispo.cec.be

———. 1999c. "Status Report on European Union Telecommunications Policy." December 22. www.ispo.cec.be

———. 1999d. *Towards a New Framework for Electronic Communications Infrastructure and Associated Services: The 1999 Communications Review.* www.ispo.cec.be

———. 2000a. *eEurope: an Information Society for All.* Lisbon. www.ispo.cec.be

———. 2000b. "Interconnection Tariffs in Member States as of 1st March 2000." www.ispo.cec.be

———. 2000c. "The Lisbon European Council—Contribution of the European Commission." www.ispo.cec.be

———. 2000d. "Proposal for a Directive of the European Parliament and of the Council Concerning the Processing of Personal Data and the Protection of Privacy in the Electronic Communications Sector." Com(2000)385.

———. 2000e. "Proposal for a Directive of the European Parliament and of the Council on Access to, and Interconnection of, Electronic Communications Networks and Services." Com(2000)384.

———. 2000f. "Proposal for a Directive of the European Parliament and of the Council on a Common Regulatory Framework for Electronic Communications Networks and Services." Com(2000)393.

———. 2000g. "Proposal for a Directive of the European Parliament and of the Council on the Authorisation of Electronic Communications Networks and Services." Com(2000)386.

———. 2000h. "Proposal for a Directive of the European Parliament and of the Council on Universal Service and Users' Rights Relating to Electronic Communications Networks and Services." Com(2000)392.

———. 2000i. "Regulation of the European Parliament and of the Council on Unbundled Access to the Local Loop." 2000/0185 (COD).

Eurostrategies–Cullen International. 1999. *The Possible Value Added of a European Regulatory Authority for Telecommunications.* Report prepared for the European Commission. www.ispo.cec.be.

Federal Communications Commission (FCC). 1996. Commercial Mobile Radio Spectrum Cap: Report and Order, WT Docket 95-59, GN Docket 90-314, released June 24, app. A.

———. 1999. *Reference Book of Rates, Price Indices, and Expenditures for Telephone Service.* Industry Analysis Division, Common Carrier Bureau (June).

———. 2000. *Telecommunications Industry Revenues: 1999.* Industry Analysis Division, Common Carrier Bureau (September).

———. 2001. *Statistics of the Long-Distance Telecommunications Industry.* Industry Analysis Division, Common Carrier Bureau (January).

Hausman, Jerry. 1997. "Valuing the Effect of Regulation on New Services in Telecommunications." *Brookings Papers on Economic Activity: Microeconomics:* 1–38.

———. 1999a. "The Effect of Sunk Costs in Telecommunication Regulation." In *Real Options: The New Investment Theory and Its Implications for Telecommunications Economics,* edited by James Alleman and Eli Noam, 191–204. Kluwer.

———. 1999b. "Regulation by TSLRIC: Economic Effects on Investment and Innovation." *Multimedia Und Recht.*

———. 1999c. "Cellular Telephone, New Products and the CPI." *Journal of Business and Economics Statistics* 17 (2): 188–94.

Hazlett, Thomas W. 1999. "Economic and Political Consequences of the 1996 Telecommunications Act." *Hastings Law Journal* 50 (August 1999):1359–94.

Huber, Peter. 1997. *Law and Disorder in Cyberspace.* Oxford University Press.

Kahn, Alfred E. 1998. *Letting Go: Deregulating the Process of Deregulation.* MSU Public Utilities Papers. Michigan State University, Institute of Public Utilities and Network Industries.

Kellogg, Michael, John Thorne, and Peter Huber. 1996. *The 1996 Telecommunications Act.* Little Brown.

Kiessling, Thomas, and Yves Blondeel. 1998. "The EU Regulatory Framework in Telecommunications: A Critical Analysis." *Telecommunications Policy* 22 (7): 571–92.

Krasnow, Erwin G., Lawrence D. Longley, and Herbert A. Terry. 1982. *The Politics of Broadcast Regulation.* St. Martin's Press.

Lehr, William, and Thomas Kiessling. 1998. "Telecommunication Regulation in the United States and Europe: The Case for a Centralised Authority." Paper presented at the annual Telecommunications Research Policy Conference, Alexandria, Va.

Leibowitz, Dennis H., and others. 2000. *The Global Wireless Communications Industry, Winter 1999–2000.* New York: Donaldson, Lufkin, and Jenrette.

Logica Consulting. 2000. *Assessment of the Leased Line Market in the European Union and the Consequences on Adaptation of the ONP Leased Line Directive.* Report prepared for the European Commission. www.ispo.cec.be

MacAvoy, Paul W. 1996. *The Failure of Antitrust and Regulation to Establish Competition in Long Distance Telephone Services.* Washington: American Enterprise Institute for Public Policy Research.

National Economic Research Associates (NERA). 1997. *Issues Associated with the Creation of a European Regulatory Authority for Telecommunications.* London.

Organization for Economic Cooperation and Development (OECD). 1999. "Building Infrastructure Capacity for Electronic Commerce—Leased Line Developments and Pricing." DSTI/ICCP/TISP 99 4/FINAL. www.oecd.org.

Pelkmans, Jacques. 1990. "Regulation and the Single Market: An Economic Perspective." In *The Completion of the Internal Market*, edited by H. Siebert, 32–60. Mohr, Tübingen.

———. 2000. "The GSM-Standard: Explaining a Success Story." Brussels: Center for European Policy Studies.

Pontiggia, Laura, and Ann Vandenbroucke. 2000. "The Impact on Competition of Differences in Telecommunications Licensing Regimes." *Communications and Strategies* 38 (2d quarter): 41–71.

Prosperetti, Luigi, Maria Cimatoribus, and Cinzia della Torre. 1999. "Experimenting with Regulatory Shortcuts in Europe." Paper presented at the annual Telecommunications Research Policy Conference.

Scherer, Joachim, ed. 2000. *Telecommunications Law in Europe.* Butterworths.

Taylor, William E., and Lester D. Taylor. 1993. "Post-Divestiture Long-Distance Competition in the United States." *American Economic Review: Papers and Proceedings* 83 (May): 185–90.

U.S. Department of Justice. 1987. *The Geodesic Network: 1987 Report on Competition in the Telephone Industry.*

About the Authors

Martin Cave is professor at Warwick Business School, Warwick University. He has written extensively on regulation of the telecommunications industry, especially in Europe. He is coauthor, with Robert Baldwin, of *Understanding Regulation* (Oxford University Press, 1999). He has advised the U.K. Office of Telecommunications (OFTEL) for many years and is a member of the U.K. Competition Commission.

Robert W. Crandall is senior fellow in the Economic Studies program at the Brookings Institution. His current research focuses on regulatory policy in the telecommunications sector, with emphasis on competition in voice, broadband, and wireless services. He is coauthor, with Leonard Waverman, of *Who Pays for Universal Service? When Telephone Subsidies Become Transparent* (Brookings, 2000); and coauthor, with Leonard Waverman, of *Talk Is Cheap: The Promise of Regulatory Reform in North American Telecommunications* (Brookings,1996). Before assuming his current position at Brookings, he was acting director, deputy director, and assistant director of the Council on Wage and Price Stability.

Thomas W. Hazlett is a resident scholar at the American Enterprise Institute for Public Policy Research. From 1984 to 2000 he was a professor at the University of California, Davis, where he taught economics and finance and directed the Program on Telecommunications Policy. He has also been a visiting scholar at the Columbia University Graduate School of Business, and in 1991–92 he served as chief economist of the Federal Communications Commission. His research has appeared in the *Journal of Financial Economics*, the *Journal of Law & Economics,* and the *Columbia Law Review*. He is a frequent contributor to *Barron's, Forbes ASAP,* and the *Wall Street Journal*.

Luigi Prosperetti is professor of industrial economics at Università degli Studi, Milano. He studied at Università Bocconi (*cum laude,* 1976) and the LSE (M.Sc.; Ph.D. 1982). He has been involved in regulatory issues in various capacities over time: as a regulator (he was a member of the Italian Price Commission in 1991–93 and advised the Italian government on the creation of independent regulators); as a board member of a regulated company (AEM, the Milan energy utility); as a researcher and consultant to major companies and government. He writes frequently for *24 ORE*, the Italian financial daily, and is currently working on a textbook on the theory and practice of regulation.

JOINT CENTER

AEI-BROOKINGS JOINT CENTER FOR REGULATORY STUDIES

Director
Robert W. Hahn

Codirector
Robert E. Litan

Fellows
Robert W. Crandall
Christopher C. DeMuth
Randall W. Lutter
Clifford M. Winston

In response to growing concerns about the impact of regulation on consumers, business, and government, the American Enterprise Institute and the Brookings Institution established the AEI-Brookings Joint Center for Regulatory Studies. The primary purpose of the center is to hold lawmakers and regulators more accountable by providing thoughtful, objective analysis of existing regulatory programs and new regulatory proposals. The Joint Center builds on AEI's and Brookings's impressive body of work over the past three decades that evaluated the economic impact of regulation and offered constructive suggestions for implementing reforms to enhance productivity and consumer welfare. The views in Joint Center publications are those of the authors and do not necessarily reflect the views of the staff, council of academic advisers, or fellows.

COUNCIL OF ACADEMIC ADVISERS